Chemistry

From the atom to beyond

PART 1: The Atom

By David J Bailey, Ph.D.

Cover Photo:

Ammonium Hexachloroplatinate crystals by Lindsy Whitlow

Forward

This set of books is designed to be a supplement to secondary chemistry curriculum. I have written this book for the high school student in mind. It is frustrating to find that text books are increasing in price and about half of the pages the students rarely glance at let alone read. I also developed this set of texts so the student can write in the text, highlight the text and annotate the text. Current textbooks are so expensive that many teachers are hesitant to recommend this.

The pictures are nice but many times extraneous. Besides the pictures, which everyone can find on the web, I have not included many problems or questions for students to answer. Also, other educational tools such as possible careers and chapter summaries are removed from this text. The internet has become a wealth of information and students can easily access this information, probably faster than I can.

There are two parts because I found that to go in depth in the topic it takes time and students have a wide range of learning pace. The high school environment leaves little time for in depth study in and after school. This is part 1, which should roughly correspond to the first year, or semester, of a chemistry class. The second part is developed to included more difficult and advanced concepts. I have remove most references to old atomic theories. I understand the importance in recognizing scientists who have made progress in the field of chemistry, but often the focus is on one group of chemists and physicists. I leave the history lesson to other texts.

I hope this helps the reader understand chemistry and the complexity of our world. Chemistry is a very interesting and often perplexing field to study.

Enjoy the book.

Acknowledgements

I would like to thank my family for being patient and supportive during this endeavor. I would also thank my editor, Becky DeJesus. She has helped me keep my prepositions straight and the text readable. Lastly, thank you for purchasing this book.

Sincerely,

Dave Bailey, Ph.D.

Table of contents

Forward iii

Acknowledgements iv

Chapter 1: Matter...1

Chapter 2: Atomic Structure – The Nucleus...................................9

Chapter 3: Atomic Stability and Radioactivity 15

Chapter 4: The Electron 21

Chapter 5: Periodic Trends.. 33

Chapter 6: Covalent Bonding 43

Chapter 7: Ionic Bonding 57

Chapter 8: Metallic Bonding, Crystal Structures, and Bond Strength 65

Chapter 9: The Mole 73

Chapter 10: Stoichiometry and Calculations............................... 82

Chapter 1: Matter

What is matter?

Everything we see, touch, smell and taste is comprised of matter. All matter from the lowly single cell organism to the molecule of sugar to the vast planetary bodies is comprised of the elements. So to begin this book, a discussion of matter is provided. Matter by definition according to Webster

> "Material substance that occupies space, has mass, and is composed predominantly of atoms consisting of protons, neutrons, and electrons, that constitutes the observable universe, and that is interconvertible with energy."[1]

This definition is more precise than most other chemistry textbooks. Let's review this definition.

Matter is a substance that occupies space and has mass. Therefore, matter must occupy space or must have a volume – no matter how large or small. All students of chemistry have experienced matter, including air. Try blowing up a balloon. Matter must also have mass, which indicates we can measure the weight of the item. The difference between mass and weight; it's actually profound.

Weight is the force by an object on a surface. The weight is defined as mass times the acceleration of gravity. Thus, the weight of the object can change if we change the location of the object from the Earth to the

---

[1] Merriam-Webster website. 10/5/15

Moon or Mars for example. However, the mass of the object remains the same. How we measure mass is by a balance, where the mass of the object is measured against a set of calibrated masses. Weight, on the other hand, is measured by setting the object on a scale and reading the amount of force applied on the scale.

Matter is comprised of atoms consisting of protons, neutrons and electrons. We will discuss this in detail in chapter 2. The last phrase of the definition, interconvertible with energy, is a nod to Einstein's theory of relativity and will be addressed later in chapter 3.

How to characterize matter?
As a chemist and a scientist, the need to categorize and sort data is strong, and matter is something that can be categorized. Of course, there are always exceptions, and it is these exceptions that have driven scientists to new discoveries. We can sort matter into two groups: Pure substances and mixtures. These groups can be sorted into subgroups as well.

Considering pure substances, we can define a pure substance as matter that cannot be separated by physical means and has uniform chemical and physical properties. Pure substances are divided into two groups: Elements and compounds. An element can be defined as an atom that has an identical number of protons between other atoms of the same element. A compound is defined as a substance containing atoms of two or more elements.

When two or more compounds are in the same container, you have a mixture. There are two major types of mixtures: homogeneous and heterogeneous. Homogeneous mixtures are those substances that have a single phase. Air that we breathe is considered a homogeneous mixture, as is ground water, flavored water, and brass. Heterogeneous mixtures are substances that have two or more phases, each component can be physically isolated easily. Examples are chocolate chip cookies, oil and vinegar salad dressing, and smoky air.

What are physical properties?

Physical properties in chemistry are those properties that do not require a change in the composition of matter to describe a change in the property. Many properties involve the observation of matter to classify the property such as color, density, malleability, and volume. Most physical properties can be observed with the obvious exception of mass, temperature, volume, and density. Temperature and volume are easy to measure, but density is a calculation.

Some of the physical properties are dependent on the amount of the substance while others are not. An extensive physical property is a physical property that depends on the amount of a substance. Examples of physical properties that are extensive are mass, volume, or total energy. An intensive property is defined as a physical property that is independent of the amount of substance. Usually described as bulk properties, examples include temperature, density, and specific heat capacity.

What is density?

Density is determined by measuring the mass of the object and dividing by its volume. The equation we use for this text is $D=m/V$, where D is the symbol for density, m is mass and V is volume. For this section, all values will be in the metric or SI system of measurements. SI system, or *systeme internationale*, of measurement is the system of measurement that scientists use to describe an object quantitatively. There are seven fundamental units of measure (Table 1.1). To help describe large or very small values, prefixes are used. (Table 1.2) Each prefix is a multiple of a thousand or a factor of three, with a few exceptions: c for centi, and d for deci-. The units for density are g/mL for solids and liquids, and g/L for gases. As a scientist and a chemist, you should be able to convert from g to mg to kg, for example.

Table 1.1: Fundamental Units of SI System

Unit of measurement	Symbol	Property
m	Meter	Length
K	Kelvin	Temperature
Cd	Candela	Light intensity
A	Ampere	Electrical current
Mol	Mole	Amount of substance

Table 1.2: Common Metric Prefixes Used in Chemistry

Prefix	name	Power of 10	Example
M	Mega	1,000,000 or 10^6	1,000,000 m = 1 Mm
k	Kilo	1000 or 10^3	1,000 L = 1 kL
d	Deci	1/10 or 0.1 or 10^{-1}	10 dL = 1 L
c	Centi	1/100 or 0.01 or 10^{-2}	100 cm = 1 m
m	Milli	1/1000 or 0.001 or 10^{-3}	1,000 mm = 1 m
μ or u	Micro	1/1000000 or 0.000001 or 10^{-6}	1,000,000 μm = 1 m
n	Nano	1/1000000000 or 0.000000001 or 10^{-9}	1,000,000,000 nL = 1 L

What are chemical properties?

Chemical properties in chemistry describe a process or quality that affects the composition of matter. Flammability, reactivity, and flash points are examples of chemical properties. Flammability is used to describe the substance's ability to react with oxygen to produce heat. Reactivity is a general term used to describe the substance's ability to react with other substances.

What are physical states?

When looking at matter, another system of classification is used to separate matter into its type of phase, which can be a gas, liquid, or a solid. A simple definition of the phases revolves around the matter's ability to have a fixed volume and/or shape. A gas is defined as matter that does not have a fixed shape or volume. A gas can fit into any shape of container and a fixed amount of gas can have any volume. A liquid is defined as having a fixed volume but not a fixed shape. 100 g of water is equivalent to 100 mL of water, but the shape of the container can be tall or short, and the liquid can fit into that container. A solid has both a fixed volume and shape. An example of this is a block of wood. The shape of the wood cannot be altered without adding or subtracting from the object and the volume is fixed.

What is a phase change?

In chemistry the phase of matter is related to a function of temperature and pressure. According to this relationship, there is a fourth phase of matter, supercritical fluids. Plasma has been taught in many science classes as the fourth phase of matter. However, in chemistry, plasma is defined to be an ionized gas – emphasis on gas. Also, if you speak to a geologist and material chemist, the phase diagrams for minerals and other materials are often complex.

Figure 1.1 is a simple phase diagram explaining the phase boundaries for any substance. The line indicates the boundary between the phases. On this line, at that temperature and pressure, each phase can coexist with each other. The point where three lines come together is called the triple point (A). This is the point all three phases can coexist with each other. Lastly, the point (B) is called the critical point. Some diagrams may not show the lines coming off this point, but above and to the right of this point is the supercritical fluid phase. Pick a point on the chart. As the temperature changes and the pressure remains constant, that point will move across a line. When that happens, you will see a phase change and there are six phases changes or transitions. Starting with solids, the transition of a solid to a gas is called sublimation

(1). This transition is easy to observe if a person obtains iodine crystals and places them in a transparent jar. If you hold the jar over low heat (your hand will do) for 10 to 15 minutes, the iodine will sublimate and the air in the jar will become slightly purple. The reverse process is deposition (2), where a gas or a vapor becomes a solid. The phase change of a substance going from a liquid to a gas is defined as boiling or vaporization (3) while the reverse is condensation (4). The last two involve the transitions between a liquid and a gas. The phase transition from a solid to a liquid is defined as melting (5). The transition of a liquid to a solid is called freezing (6).

Figure 1.1: Generic Phase Diagram

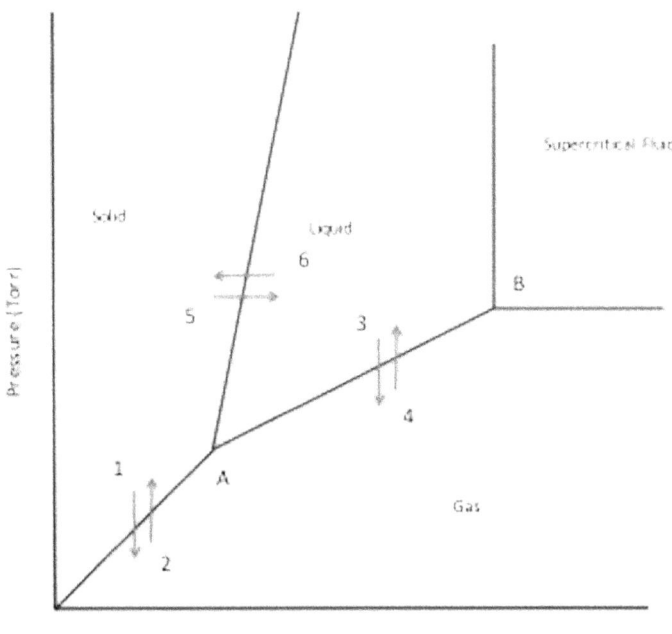

Is there a difference between a chemical reaction and a change in the physical property?

A change in a physical property, due to either a change in temperature, pressure, or force, is a reversible condition. For example, when you boil water, the liquid water becomes a gas and when the temperature of the steam decreases, you will have liquid water again. A chemical reaction is observed when an irreversible change in a physical property occurs or a phase transition is observed when the conditions specify it should not occur. For example, if we add zinc to a solution of hydrochloric acid, we observe bubbles forming. Even with the increase in the temperature of the solution, it is not enough to cause water to boil. Therefore, we can assume that a chemical reaction has occurred, and we can capture the gas and test it.

Chapter 2: Atomic Structure – The Nucleus

What are atoms?

In the previous chapter, we focused on the general properties of matter. Continuing our study into chemistry, the smallest part of matter is next: the atom. The structure of the atom is simple and sophisticated. Dalton actually hypothesized the existence of the atom and its effects in the 1700's. His postulates became the building blocks of chemistry and are discussed below.

What are Dalton's Laws?

Dalton defined elements comprised of small, indivisible particles called atoms. However, we do know that atoms are comprised of protons, neutrons, and electrons and can be divided. In this chapter we will discuss the nucleus, and later in the book, the electrons. Dalton's first postulate can be stated as follows:

> "the smallest, identifiable part of the element is called the atom, which consists of protons, neutrons and electrons."

The second postulate is that every atom of an element is identical in every way: mass, size, physical, and chemical properties. Atoms of other elements are different. What we know now is that atoms of a single element may differ in term of mass, but identical in everything else. Atoms of a single element that have a different mass are called isotopes of each other. This postulate of Dalton's atomic theory can be restated:

> "Atoms that have atoms of identical number of proton are identical in physical and chemical properties, mass, and size.

There are elements that contain atoms with various masses and these atoms are defined to be isotopes of each other."

The third postulate is that atoms cannot be created or destroyed. Since 1945, when the atomic weapons were developed, we know that is not the case. We use Einstein's equation of $E=mc^2$ to develop relationships between mass and energy. This postulate can be rewritten as follows:
"Chemically, atoms cannot be created or destroyed."

The fourth postulate of Dalton's atomic theory is that atoms react to form compounds that are a whole number ratio of the elements in the substance. This statement is still true today and has not been changed. Now, if you continue to study chemistry, there are a few exceptions to this postulate. The exceptions are rare and the explanations will be discussed in another text.

What are the parts of the atom?
The atom can be divided into two areas: the nucleus in the center of the atom, and the rest that consists of empty space and electrons. In the early 1900s Rutherford postulated that the atomic structure consisted of a positively charged nucleus, or center, surrounded by electrons. He came to this conclusion after conducting his now famous bombardment of gold foil using alpha particles. The parts of the nucleus weren't realized until later. In this text, the explanation will stop at the proton, neutron and electron. Other sub atomic particles will be discussed in a different text. Although the electron was the first subatomic particle to be discovered, a detailed explanation of the electron and its chemistry is discussed in the chapter 4.

What is a neutron?

Chadwick discovered the neutron by his bombardment of beryllium by alpha particles. Neither magnetic or electric fields could deflect the resulting beam that was produced. Thus, the particle had to be neutral. Since the analysis of the beryllium showed traces of carbon, the result had to be a neutral particle coming from the nucleus.

What is the positive charge in the nucleus?

Rutherford isolated the proton by the irradiation of nitrogen by alpha particles. The result was the formation of oxygen-15 and a proton. A chemist from England in the 1800's, William Prout, hypothesized that all the elements were made from hydrogen and the discovery of the proton provided some justification of this hypothesis.

What is the mass of the proton and neutron?

Determining the mass of these particles was a bit more challenging. Realize that the protons identify the element and provide some of the mass. Neutrons provide the rest of the mass of the nucleus. Neither have any bearing on the chemistry of the element; the electrons and the electronic orbitals help dictate the chemistry of the elements. At first approximation, we assume that the mass of a proton and a neutron are equivalent and are equal to 1 amu. AMU is defined as a unified atomic mass unit (u) and is also known as a Dalton (Da). The AMU is the non-SI system of measurement and is equivalent to 1.66×10^{-27} kg. Dalton is another unit of mass, equivalent to an AMU. Biochemists and biologists often use Dalton in their work. In this text, I have decided to use the term Dalton for atomic masses.

Table 2.1 – Mass of Subatomic Particles

Electron	0.0005486 Da	9.107×10^{-31} kg
Proton	1.0073 Da	1.672×10^{-27} kg
Neutron	1.0087 Da	1.674×10^{-27} kg

What are atomic symbols?

Elements are the simplest form of matter. Containing atoms from one element, the elements can combine to form simple and complex molecules and materials. Each element has a name. Some were named during the Greek and Roman era of discovery while most were named during the period of discovery in the 1800s.

The elemental symbol consists of one or two letters; the first letter is always capitalized. Commonly, the first letter of the symbol corresponds to the first letter of the name – but there are a few significant differences with elements named prior to 1800's. All but 12 elemental symbols have two letters. The twelve elements that only have a one-letter symbol are usually those that are well known. Hydrogen (H) is the only element that has more than one elemental symbol. ^{2}H is also written as D for deuterium and ^{3}H is also written as T for tritium.

How are the elements named?

The elements are named due to its color, where it is from, the person of note, or a variety of other reasons. The elements that were known to the Greeks and Romans have Latin names and the symbols are derived from those names. Table 2.2 provides a list of these elements and the corresponding name and symbol. There are several books that explain the history of the names. The "CRC Handbook of Chemistry and Physics" is one of the official reference books in chemistry and it has a short but detailed section on history of each element.

Table 2.2: Names and Symbols of Elements Derived from Latin

English	Latin	symbol
Sodium	natrium	(Na)
Potassium	kalium	(K)
Silver	argentium	(Ag)
Gold	aurium	(Au)
Iron	ferrium	(Fe)
Copper	cuprium	(Cu)
Lead	plubium	(Pb)
Mercury	hydrargyrum	(Hg)

What information is present in chemical symbols?

Looking at the chemical symbols again, Figure 2.1 describes the information the chemical symbol can contain. There are four places where chemists usually place numbers to provide information about the element. Lower left is where the atomic number is placed. The atomic number must coincide with the elemental symbol and is equivalent to the number of protons in the atom. Upper left is the mass of the isotope. This mass is the sum of the protons and neutrons in the nucleus. The mass of the electrons are ignored for this text, because the effect of the electron mass is not significant unless we are in the extremely heavy metals, usually atomic number 90 and higher. The upper right is the charge on the atom or molecule. If the value is negative, the atom has gained electrons. If the value is positive, the atom has lost electrons. Please note that this is contrary to most numbering conventions. However, remember that electrons are negatively charged, and if you increase the number of negative charges to positive charges, the overall charge becomes more negative. The lower right of the symbol is used to convey the number of atoms of that element in the molecule or material.

Figure 2.1: Generic Chemical Symbol

$$_{z}^{m}X_{n}^{c}$$

Where m is the atom's mass in Daltons (Da), z is the atomic number, c is the charge on the atom and n is the number of atoms in a molecule.

What is an isotope?

Isotopes are defined as an atom of the element that has a different number of neutrons, thus a different mass, than the other atoms in the element. For example, carbon has three isotopes, ^{12}C or carbon-12, ^{13}C or carbon-13, and ^{14}C or carbon-14. Some of the isotopes are stable, but some will undergo a transformation to another element.

Chapter 3: Atomic Stability and Radioactivity

How can the nucleus contain both protons and neutrons?

All atoms contain energy in one form or another. The form of energy can be transferred or changed to another form easily. This book will go in depth with energy and its transference later in part 2. However, we will describe the energy in the nucleus below.

The nucleus has kinetic energy in the form of spinning. Even as a solid, the nucleus will spin at a specific rate. Looking closer at the nucleus, one can observe that the protons will spin as well; this can be deduced from the fact that the protons have a magnetic moment. A moving charged particle will create a magnetic field. We can probably assume that the neutrons spin as well. The role of the protons is to provide mass, charge, and the identity of the atom. The role of the neutrons is to act as a charge buffer between the protons in the nucleus and to provide mass.

The nucleus also contains a force that binds the protons and neutrons together called the binding energy. This binding energy is important since the nucleus needs some force to keep the positive protons in the center of the atom.

Binding energy can be calculated by summing the mass of protons and neutrons in the isotope. Then, by subtracting the actual mass from the calculated mass, the mass defect is calculated. This difference is used to calculate the binding energy by using Einstein's relationship of $E=mc^2$. This can be considered a type of potential energy, which can be transformed back to mass or emitted as gamma radiation if the nucleus is unstable.

What is meant by atomic stability?

The nucleus may undergo a transformation; if the nucleus is transformed scientists characterize it as unstable. The stability of an atom's nucleus varies depending on the ratio of neutrons to protons and the number of protons in the atom. Some elements have many stable isotopes while a few contain many unstable isotopes.

Many elements contain isotopes with different mass or number of protons and neutrons in the nucleus. An element that has two distinct atoms with different number of neutrons is described to have isotopes. An isotope is an atom that has the same number of protons but different number of neutrons. There is a stability of the atom with a ratio of neutrons to protons, which is 1:1 for light elements and increasing to 1.5:1 for elements containing less than 82 protons (lead). All the nuclei are unstable for elements with more than 82 protons. The term we use to describe this condition is radioactive.

What is radioactive decay?

A radioactive element will decompose to another element by one of the pathways described below or a combination of one pathway plus gamma radiation. The different types of radioactive emission are alpha, beta, positron, gamma, and electron capture. There are occasions that gamma radiation is the only emission from the nucleus, but more often gamma radiation occurs with one of the other radiation emissions.

How to balance radioactive reactions?

To balance radioactive reactions, there are several assumptions to keep in mind. One, mass is conserved. This means that the total mass we begin with is equivalent to the total mass of the products. Two, the atomic number is conserved; the total number of protons is equal on both sides of the equation. At this time do not worry about the total charge of the radioactive decay. It is important, but for this class we will ignore the charges in this chapter.

What is alpha decay?

Occurring in elements with an atomic number greater than 82, alpha radiation is the emission of a helium atom from the nucleus of the atom. The alpha particle has a charge of +2, so when it hits a substance, the atoms in the material are ionized and become positive as well. The chemical symbol for an alpha particle is $^{4}_{2}\alpha$ or $^{4}_{2}He$. For the nuclear reaction written in this chapter, the charge on the atoms is omitted. The reader should realize that any excess electrons are emitted after the decay occurs.

Example of an alpha decay:

$$^{228}_{90}Th \rightarrow {}^{224}_{88}Ra + {}^{4}_{2}\alpha$$

Note that the total mass right of the arrow is equal to the mass on the left. The sum of the protons right of the arrow is equal to the number of protons to the left. Therefore, the equation is balanced.

What is beta decay?

Beta decay occurs when an isotope has too many neutrons compared to protons. When this decay occurs, a high-energy electron is emitted, called a beta particle in this case, and the atom gains a proton. The mass of the atom remains the same. The chemical symbol for a high-energy electron or beta particle is either $_{-1}^{0}\beta$ or $_{-1}^{0}e$. An example of beta decay is shown below:

$$_{16}^{35}S \rightarrow _{17}^{35}Cl + _{-1}^{0}\beta$$

What is positron decay?

Positron decay occurs when the isotope has too many protons compared to neutrons and the isotope has an atomic number less than 82. The mass of the positron particle is identical to the electron, but it has a positive charge. This particle will react with an electron to annihilate the electron and produce gamma radiation. Therefore, positron radiation can be called antimatter. The nucleus will transform a proton to a neutron and the mass will not change. An example of a positron emission is shown below:

$$_{19}^{40}K \rightarrow _{18}^{40}Ar + _{+1}^{0}\beta$$

The symbol for a positron particle is either $_{+1}^{0}\beta$ or $_{+1}^{0}e$

What is electron capture decay or process?

Electron capture is a unique type of radioactive decay. Plainly speaking, the electron is captured by the nucleus of the atom and reacts with a proton to produce gamma radiation and a nucleus with one additional neutron. The mass does not change with this transformation. The chemical symbol for electron capture process is either *EC* for electron capture or an electron $_{-1}e$. The mass of the nucleus remains the same. An example of electron capture is shown below:

$$_{27}^{57}Co + EC \rightarrow _{26}^{57}Fe + \gamma$$

With this example, cobalt has 30 neutrons and iron has 31 neutrons.

When does gamma radiation occur?

Gamma radiation is a high-energy photon of light. It is the result of the nucleus changing its energy state. Gamma radiation is generally produced from the resulting decay of the nucleus, although there are some radioactive processes that result in a gamma emission.

What is a fission reaction?

Fission is a radioactive process that involves the splitting of the nucleus, either spontaneously or by the absorption of a high-energy neutron. The symbol for a neutron is ^1_0n. The result is gamma radiation and two or more nuclei that are significantly smaller than the original nucleus. A classic example of nuclear fission is shown below:

$$^1_0n + ^{235}_{92}U \rightarrow ^{139}_{56}Ba + ^{94}_{36}Kr + 3\,^1_0n$$

What is a transmutation reaction?

A transmutation reaction occurs when a target atom is bombarded or collided with another atom or subatomic particle. The result is an atom with a higher mass and atomic number than the target atom with an emission of gamma radiation and other particles. To balance these types of reactions, make sure that the total atomic number is equivalent on each side of the arrow and the total mass is the same as well. In the reaction below, californium is the target atom:

$$^{252}_{98}Cf + ^{10}_5B \rightarrow ^{256}_{103}Lr + 6\,^1_0n$$

What is a fusion reaction?

Fusion is a radioactive process that involves the combination of two nuclei to form a larger nucleus and the emission of gamma radiation. The difference between fusion and transmutation is that fusion involves two nuclei that are similar, if not identical, in size, while transmutation involves two nuclei of different sizes. Also, the transmutation reaction usually has two nuclei as products compared to the fusion reaction that has one nuclei as a product. An example of a fusion reaction is the fusion of two hydrogen atoms to form a deuterium nucleus and a positron particle:

$$^1_1H + ^1_1H \rightarrow ^2_1D + ^0_1\beta$$

What is meant by half-life?

Half-life is the time it takes one half of a sample of radioactive material to decay. For example, if you have 10g of carbon-13, it will take over 5000 years for 5 g of it to decay into nitrogen. Detailed explanation and examples can be found in Part 2, Chapter 3 – Kinetics.

Chapter 4: The Electron

Why spend some much time studying the electron?

The electron has a very complex structure and role for the atom. The energy of the electron in the atom dictates the reactivity of the element. Scientists have been able to measure the energy and the mass of the electron with great precision; however, the actual position of the electron will elude scientists due to Heisenberg's postulate. However, with quantum mechanics, scientists can determine the probability of finding an electron is a volume of space around the atom. This probability also assists in the reactivity of the atom. This chapter will discuss the results of quantum mechanics.

Why does an atom feel solid?

The hardness of the atom is an illusion produced by the electrons travelling at very high speed. As an approximation, if we assume that the electron travels at 0.01 times the speed of light, this electron will have traveled 3.00×10^6 m in one second or would have made 2.25×10^8 trips around an atom. This approximation is only valid when we look at normal physics. If we consider quantum mechanics, the results will be different. Therefore, it is very difficult to determine the position of the electron with our technology today.

Why is the electron hard to find?

The electron's position in the atom cannot be determined; this is due to a theorem called Heisenberg's Uncertainty Principle. There is a limit to how small we can detect or measure a moving object, and the higher the precision in the measurement, the less likely we can determine the other values. The Heisenberg's Uncertainty Principle can be described as the precision in measuring the position of the object times the precision in measuring the momentum of the object, momentum is mass times velocity, is greater than 5.265×10^{-35} kg*m^2/s. The actual equation is $\Delta x(mv) = 5.265 \times 10^{-35}$ kg*m^2/s Since we know to a high precision the energy of an electron and the velocity, we will have difficulty determining the precise location of the electron.

Who discovered the electron?

Thompson discovered the electron in the 1900s by investigating the discharge from a Crook's tube or a cathode ray tube. This discharge was determined to be influence by a magnetic field and an electric field. The beam was deflected towards a positive plate, hence the negative connotation, and the charge to mass ratio was determined by measuring the deflection caused by magnetic fields. The electron was the easiest particle to remove from the atom.

Millikan calculated electron's mass in 1909 and 1910. They are very small and readily able to leave the atom, which enable the elements to react. Other physicists, including Schrödinger and de Broglie, hypothesized that the electrons move around the nucleus of an atom in a wave-like state and are held by the nucleus in an electrostatic attraction with the energy of the electrons having specific values.

Thus, the field of quantum mechanics came into being. The results of quantum mechanics are very simple to explain – but it can be very complex at the same time.

One result of the quantum mechanical functions is to calculate the probability of finding an electron with a specific energy. This translates to a volume of space around the nucleus where we predict the electron to be 95% of the time. Each energy level and orbital have a specific shape and maximum distance associated with it.

What are electron orbitals?

Electron orbitals are a volume of space where the probability of the electron can be found. This probability is determined by calculations from quantum mechanics. In this chapter, the results of these calculations are described. An orbital has a unique set of n, l, and m quantum numbers. Only two electrons can have that set of numbers, one electron having ½ spin and the other having - ½ spin. Thus, each electron should have a unique set of quantum numbers and each orbital can contain two electrons. Figures 4.1 and 4.2 shows the picture of the s and p orbitals.

Figure 4.1: Diagram of s-Orbital

s-orbital

Figure 4.2: Diagrams of p-Orbitals

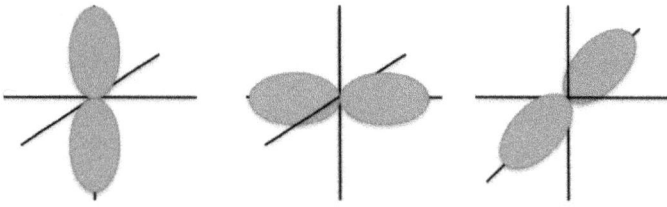

p-orbitals (z, x, and y)

What are quantum numbers?

The set of quantum numbers are described in this text as (n,l,m,s) where n is the principle energy level, l is the angular momentum of the orbital, m is the magnetic quantum number, and s describes the electron's spin. Each electron in the atom will have a unique quantum number, which will describe the energy level of the electron and ultimately the region of space around the atom where we can expect to find it at a 95% probability.

The first number, the principle energy level, is the primary energy level of the electron. The n number ranges from 1, 2, 3, … but usually the numbers stop at 8 until scientists can create more elements. The first energy level only has one orbital, but the second energy level has two, the third three and so on.

The angular moment quantum number (l) describes the angular moment associated with the electron in the orbital. The values of l ranges from 0 to n-1. So, if n=2, l = 0 or 1. Now the l values are assigned letters to describe the orbitals, see Table 4.1 for details.

Table 4.1: Quantum numbers

L number	Orbital
l = 0	s – orbital
l = 1	p – orbital
l = 2	d – orbital
l = 3	f – orbital
l = 4	g – orbital (theoretical only)

The magnetic quantum numbers (m) are whole numbers that range from −l to l including zero. These orbitals are equivalent in energy and are detected when a magnetic field is applied. The magnetic quantum number dictates the number of orbitals that have the same energy, called degenerate orbitals. For example, if l = 1, this is the p-orbitals and m = -1, 0, and 1, giving the p-orbitals a degeneracy of 3. In other words, there are three p orbitals for every energy level. Table 4.2 shows the number of degenerate orbitals for each type of orbital.

Table 4.2: Orbital and number of degenerate orbitals

Orbital	Number of orbitals
s	1
p	3
d	5
f	7

The last quantum number is the electron spin quantum number (s). It will have a value of either - ½ or + ½. This quantum number will enable the chemist to determine the strength of the element's magnetism as well.

What are orbital energy diagrams?

As a chemist, I have been trained to view electrons in the terms of an energy graph, where the orbitals start at the bottom and increase in energy as they travel up the page. As the atomic number increases, the energy of the lowest levels decrease and the atom adds more orbitals. Like pouring a glass of water, the chemist must fill the energy chart with electrons from the bottom up – this is called *aufbau* principle. Once we start with the p orbitals, the orbital is split into 3 equivalent energy orbitals (5 degenerate orbitals for the d orbitals, and 7 degenerate orbitals for the f orbitals). So, the electron fills up the orbitals without pairing up first and all of the electrons will have similar spins. As more electrons are added, the electrons pair up but have opposite spins.

Following are the energy diagrams for hydrogen (Figure 4.3), carbon (Figure 4.4), and vanadium (Figure 4.5). Looking at hydrogen, just the lone electron in it will fall to the bottom of the energy diagram. An apt description is filling a water glass; the water will fill from the bottom up. For carbon, the electrons will fill the empty degenerate orbitals first with positive spin. The positive spin is first by convention. Also by convention is filling the degenerate orbitals from left to right. Vanadium is different because the 4s orbital is filled before the 3d. This is due to electron-electron repulsion and distance away from the electron. We know this is accurate due to vanadium's emission spectrum.

Figure 4.3: Orbital diagram for hydrogen

3p ——— ——— ———

3s ———

2p ——— ——— ———

2s ———

1s ⊥ Hydrogen

Figure 4.4: Orbital diagram for carbon

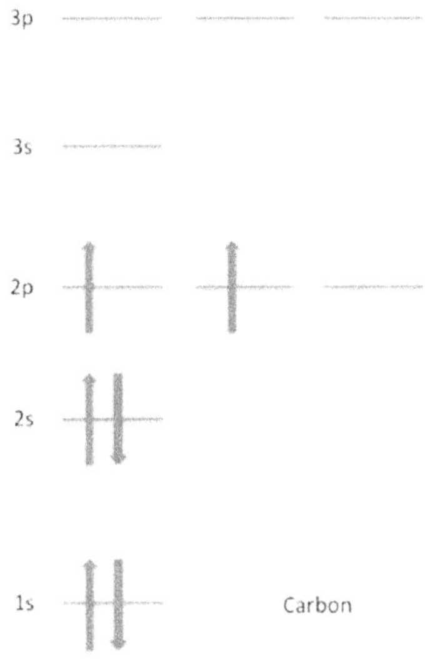

3p ——— ——— ———

3s ———

2p ⊥ ⊥ ———

2s ⊥⊥

1s ⊥⊥ Carbon

27

Figure 4.5: Orbital diagram for vanadium

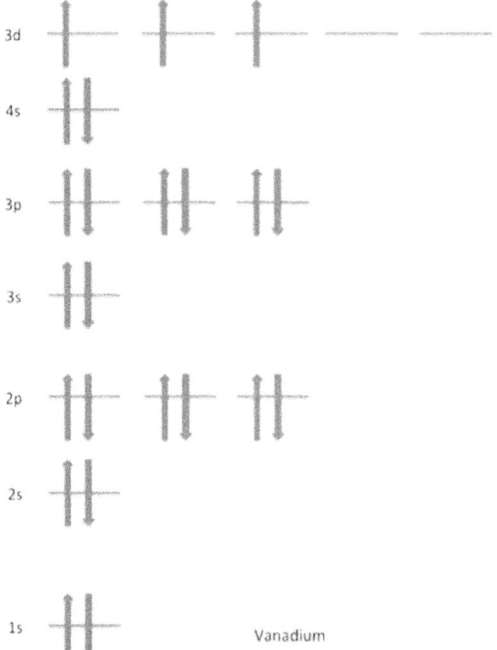

Vanadium

What is the electron configuration of the element?

The electron configuration is a brief list of the electrons and the orbitals in which the electrons reside. There are two ways to write the configurations, the long and short version. To write the long version of the electron configurations of an element, start with the number of electrons the element has. Then, starting with the 1s orbital, add electrons two at a time moving to the right.

How do you write electron configurations?

General notation of electron configuration looks like this: $1s^2$ – where the coefficient is the energy level of the orbital, s is the orbital, and the superscript or exponent is the number of electrons in the orbital. The following two methods to list the electron configurations are very compact compared to the energy diagrams, but the energy diagrams are useful when determining the number of unpaired electron in the atom or when determining other energy changes in the atom.

Hydrogen: $1s^1$

Carbon: $1s^2\, 2s^2\, 2p^2$

Vanadium: $1s^2\, 2s^2\, 2p^6\, 3s^2\, 3p^6\, 4s^2\, 3d^3$

How did we relate light to electrons?

When scientists were looking at the emission of light by the elements, they found a complex set of photons being emitted at various wavelengths. Looking at the hydrogen atom, they found several distinct wavelengths. When they tried to produce light from H^+, they couldn't. This was a hint that the electrons were involved in the production of light. A Swiss mathematician, Balmer, found that the wavelengths of light can be mathematically modeled and this model worked for a one electron atom. Balmer described the process as the atom gains energy, the electron is excited, gains the energy, and the electron moves up an energy level or more. These excited energy levels are virtual or unstable. Then the electron relaxes and loses energy in the form of a photon, and the electron is returned to a lower energy level.

Can this be explained using atomic orbitals?

So, if we use the atomic orbital theory, we have a set of electrons belonging to our element, and taking the highest energetic electron, it is excited to another atomic orbital. This orbital has to be in a higher energy level and a different type of orbital. For example, an excitation from a 2s to a 3s orbital is not allowed.   An example using orbital diagrams is shown in Figure 4.6, using carbon as an example.

Figure 4.6: Excitation of an electron in a carbon atom

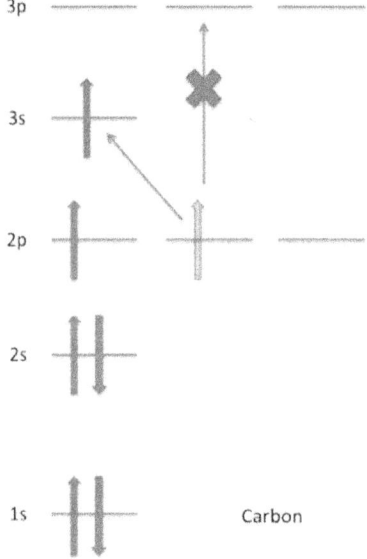

As you can see, the electron in the highest occupied orbital is the only one excited by energy and the electron will move to a virtual orbital, usually the 3s. Then the electron will lose the excess energy, in the form of a photon of light, and return to its initial orbital.

What is a photon of light to a chemist?

Light is a parcel of light that acts as a particle and a wave. The photon has a little mass, but is mostly energy. It has an electric and a magnetic component as it moves through a vacuum. The photon has a finite length and can knock electrons from atoms. In this part of the chapter, we will focus on the properties of light.

How can light act as a wave?

The wave properties of light have been studied extensively. Light and the associated colors that we see are defined as being part of the electromagnetic spectrum. Light wave has two components, an electrical component that travels through space as a sine wave, and a magnetic component that also travels through space as a sine wave. The electrical and magnetic components are right angles or perpendicular to each other. Electromagnetic waves are the only type of wave that can travel through a vacuum. All other waves, especially pressure waves such as sound, need matter to travel from one location to another.

Light can be reflected like a sound wave, it can also be refracted through a transparent material. Light can be used to make interference patterns, which are bands of constructive and destructive interference of light. Unlike sound, light can decrease its velocity when travelling through a transparent and dense substance.

Einstein developed a neat little experiment that showed to the world that light can act as a particle too. We now know that a photon of light has a size; it is finite in length. The trace the photon of light makes as it travels through space is called a wave train.

What are the Equations Associated with Light?

Chemists use several equations when working with light. There is a relationship between the velocity of light in a vacuum and its frequency and wavelength.

c=vλ

Where c = 2.98x108 m/s, λ is the wavelength of light in meters, and v is the frequency in Hz or 1/s. The energy of a photon of light is calculated by the following equation.

E=hv

Where h is Planck's constant – $6.626x10^{-34}$ J*s, and E is the energy of the photon.

Common properties of light

This means that as the frequency of light increases, the energy of the photon increases and the wavelength decreases. Also, when light is travelling through a different material than a vacuum, the velocity changes and the wavelength changes. The frequency does not change. This is true only for light and not for any other system of waves.

What is the Electromagnetic spectrum?

The electromagnetic spectrum is the range of wavelengths, frequencies, or energies that a wave can be emitted and the classification of these waves. The electromagnetic wave with the highest energy is gamma radiation. The source of gamma radiation is the nucleus of an atom. Then x-rays are next followed by ultraviolet radiation down to radio waves. The range of frequencies that the human eye can see is very small compared to the entire spectrum.

Chapter 5: Periodic Trends

What is a periodic table?

A simple periodic table contains the element's name and symbol, the atomic number and the average atomic mass of the element. Present on the table is a black line that looks like a staircase going from boron to polonium. This line breaks up the periodic table between the metals and nonmetals. Elements on the broader side of this line are called the metalloids or semimetals. There is also a color scheme for the elements. At room temperature (25 °C), if the element is a solid, the symbol is black. If the element is a liquid at room temperature, then the color of the symbol is blue. A red symbol indicates the element is a gas at room temperature. If the symbol is white or an outline, the element is man-made or manufactured.

Who developed the first periodic table?

Mendeleev developed the periodic table in its current form in the 1868. The original table listed the elements by their average atomic weights from the lowest to the heaviest. The table was also separated by chemistry of related elements. Mendeleev noticed that there were some gaps in the table that he hypothesized that there should be an element that was undiscovered. Later, Mosley suggested that using atomic number is an accurate method to sort the elements and explains some of the discrepancies. There are several parts of the periodic table that chemists use consistently.

Is the shape of the table important?

The shape of the periodic table helps with determining electron configuration and reactivity. The columns are labeled with the group number. For modern periodic tables, the numbers should range from 1 to 18 from left to right. There may be another set of numbers ranging from 1 to 8 with a letter A or B. The location of the A and B is different depending on your country. In the United States, the letter A was used to identify the main group elements and the column number indicated the number of valence electrons the element has. Column IV A indicates that those elements have 4 valence electrons and is a main group element.

The letter B was used to identify the transition metals. The column numbers start with IIIB because those elements can form a +3 cation. Three columns are identified as VIII or VIIIB. These metals can lose up to eight electrons. The last columns are labeled IB and IIB. These elements, due to their electron configuration, can lose up to two electrons. IB can lose one electron while IIB can lose two electrons. These elements are similar to alkali and alkali earth metals because under most conditions they have one charge to the element.

The first two columns are referred to as the s block elements because the last electron is in the s orbital. The traditional names for these columns are the Alkali (column 1) and the Alkali Earth metals (column 2). Columns 13 – 18 are referred as the p– block elements. These elements have the last electron in the p orbital. Most of these elements are considered non-metals, especially in the first three rows. Column 18 are the noble gases, which indicate that they do not react, and column 17 elements are called halogens, which mean salt makers. Column 16 elements are called the chalogens, which means chalk makers, and column 15 elements are called pnicogens.

Columns 3 – 12 are defined as the transition metals and have an electron configuration that ends with s and d orbitals. These elements will give up an electron and have many stable ionic states. These elements will be discussed in later chapters. There is one column that is named, the column containing gold, silver, and copper; the coinage metals.

Do the rows of the periodic table mean anything?

The row of the table is the highest principle energy level for the element. For example, calcium has electrons filling up to the third energy level. The first row only has two elements, Hydrogen and Helium. This makes sense if you considered the first energy level has the s-orbital only. The second energy level has an s and three p orbitals, for eight electrons. There are eight elements in the second row. Now, the third row should have 18 elements in it not the eight it has. But remember Chapter 4, the 3 d orbitals are filled after the 4 s. That is shown in the period table, where the first row of transition metals are located in the fourth row and the second column.

What are valence electrons?

Valence electrons are those electrons that reside in the highest energy level, usually in the s and p orbitals. Elements that have a full s and p orbital are very unreactive under normal conditions.

How do atoms become charged?

Atoms can become charged by two different methods, either by losing electrons or by gaining electrons. Metals tend to lose electrons and nonmetals tend to gain electrons. To lose electrons, the valence electrons absorb energy from the environment. The electron is then ionized and is free from the atom. The atom becomes positively charged and is called a cation. The atom will lose as many electrons as it has before it reaches a full s and p orbital combination. Figure 5-1 explains this process visually.

Figure 5-1: Ionization of sodium atom

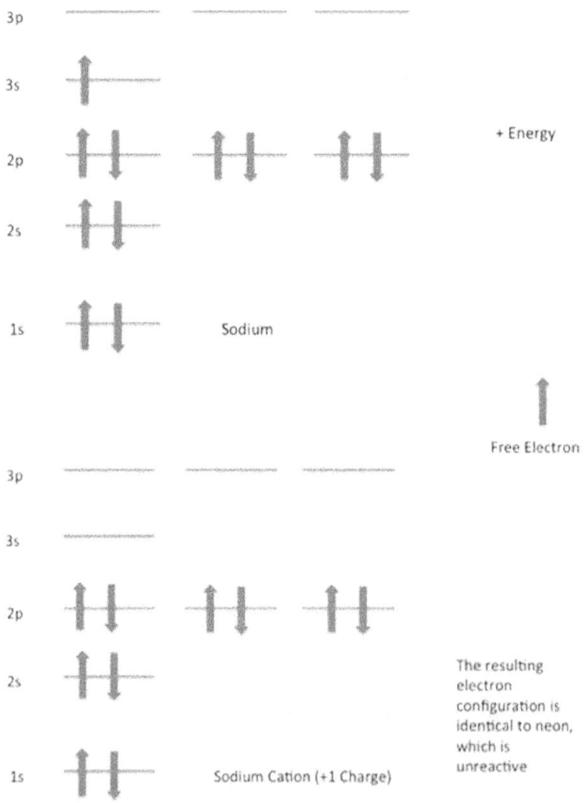

3p

3s

2p   + Energy

2s

1s   Sodium

                                          Free Electron

3p

3s

2p

2s   The resulting
                                        electron
                                        configuration is
                                        identical to neon,
                                        which is
1s   Sodium Cation (+1 Charge)   unreactive

When an atom absorbs an electron, it absorbs as many as it needs to reach a stable s and p configuration. When it reaches that state, the atom will release energy as the orbitals lose energy because of the stable configuration it has reached. Figure 5-2 explains this process visually.

Figure 5-2: Ionization of oxygen atom

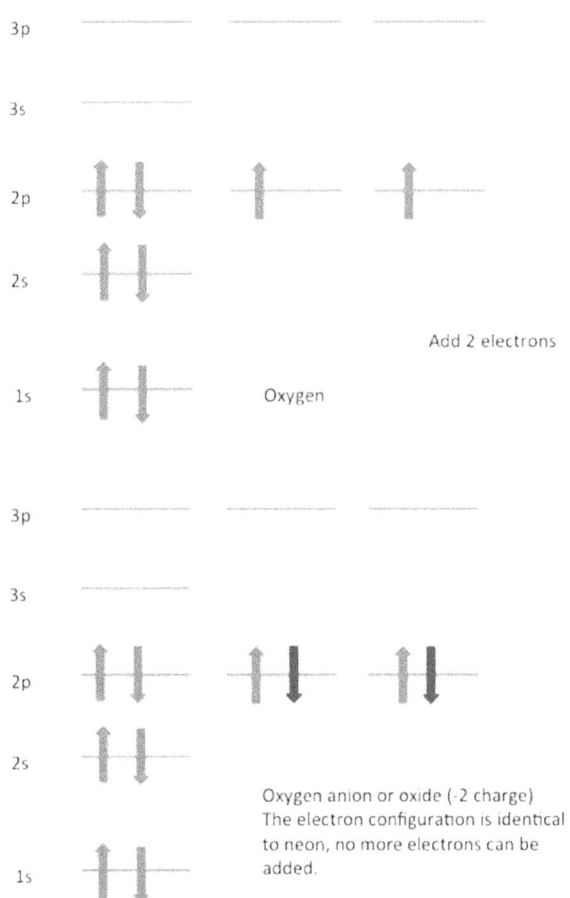

Add 2 electrons

Oxygen

Oxygen anion or oxide (-2 charge)
The electron configuration is identical to neon, no more electrons can be added.

What is ionization energy?

Ionization energy is the amount of energy needed to remove the highest energetic electron from the atom. (I usually describe this as removing the last electron from the atom.) For the transition metals, the last electron is in the s orbital. For example, the first electron to be removed in the ionization of Iron is the electron in the 4s orbital. This happens because the 4s energy level is higher than the 3d, even though the 4s orbital is filled before the 3d.

If electrons are removed from the atoms, the ionization energy increases. This is true for the valence electrons. Now, once you have removed all the electrons from the valence shell, the ionization energy increases substantially so it is very difficult to remove the electron.

What is Electron Affinity?

Electron affinity is the amount of energy that is released when an electron is absorbed by an atom. Since the atom is losing energy, the values of electron affinity are negative. This can be explained by the lowering of the orbital energy when all the electrons in the p orbitals are paired up and the observed nuclear charge decreases due to increased shielding. In this text, we will focus on the general trend of the anions and leave any exceptions to this trend to college level texts.

Why is there only one mass for the elements?

Earlier in the book, we discussed the isotope mass as the sum of the proton and neutrons in the nucleus. Since many elements have more than one isotope, the periodic table lists the average mass of the element as the atomic mass. This average is calculated using the weighted average method, which means that the mass of the isotopes is multiplied by its abundance then summed. For example, carbon has three naturally occurring isotopes, ^{12}C, ^{13}C, and ^{14}C. Carbon-12 has an abundance of 98.93%, carbon-13 has an abundance of 1.07%, and carbon-14 has an abundance of less than 0.001%. Take the mass of each isotope and multiply it by the abundance, then add the totals together. The resultant is the average atomic mass of the element. The general trend of the average atomic mass is that it increases from left to right and increases going down the column. There are a few exceptions to this trend, most notably argon and potassium, and cobalt and nickel. The unit for the average atomic mass can be either Da or g/mol. Which one to use will depend on whether you are looking at the average mass for one atom or the average mass for a mole of that element. The magnitude is the identical for each unit.

What is electronegativity?

The ability of the atom to attract electrons to the atom is called electronegativity.   This property of the atom helps the chemist determine if the bond is ionic, covalent, or polar covalent. The general trend is that the electronegativity increases as you travel from left to right in the row, and it decreases as you go down the column. The lowest electronegative value resides with francium. The highest is fluorine then oxygen, which is an important fact to remember when we start with reduction and oxidation reactions later in the text.

What is the atomic radius?

The atomic radius is the distance between the nuclei of two atoms bound together. The general trend is that the radius increases as the elements go down the column and increase going from left to right. However, the trend going from left to right is affected by the effective nuclear charge, which cause the radii to decrease after the second column and the sixteenth column. The first is due to a change in the energy levels – going from the s orbital to the p orbital. The second is due to the effect of similar electron spin in the orbital.

What is shielding?

Shielding is when the inner most electrons hide or reduce the positive charge of the nucleus from the perspective of the outer most electrons. The effect of this shielding is that the effective nuclear charge increases across the periodic table, depending on the electron configuration. Looking at Figure 5.3, the energy diagram for silicon, notice that there are two electrons in the highest energetic orbital, the 3p. These electrons are not shielding the nucleus from each other. The electrons in the other orbitals are shielding the 3p electrons from the nucleus. The amount of shielding will depend on the orbital and its energy. Electrons in the same set of orbitals do not shield the nucleus from each other; they must be at a lower energy level.

Figure 5.3: Orbital diagram of silicon

What is the ionic radii?

The ionic radius is the radius of the ion at its common oxidation state, either a cation or an anion. The cations are smaller than the neutral atom, and the anion is larger than the neutral atom. A deeper review of this trend is that as the charges increase in the positive direction, the radii become smaller. Conversely, if the charge becomes more negative, the ionic radii increase.

Is there a trend for melting points?

Yes and no. There is a trend going across the periodic table. Looking at the solids in row 3, the melting points increase until the nonmetals are reached. The next metals increase in melting points, but are lower than the previous metals. Then the gases have lower melting and boiling points, of course. The changes in trends are due to a combination of electron configuration and electron shielding. However, going down the column, the melting points should increase and they do if the crystal structure remains the same. Changes in the trend are due to the change in the crystal structure.

Is there a trend for boiling points?

Going down a column, the boiling points increase due to size. However, there are a few exceptions, due to electron configurations. Unfortunately, there is no trend going left to right along a row.

Is there a trend for density?

Going down a column, the density will increase. This is especially true for the transition metals and solids in the nonmetals section of the periodic table. Remember, for gases the density units are g/L, which is smaller than the g/mL for the solids and liquid. Going left to right along the row, the density increases until column 9, then it decreases.

Chapter 6: Covalent Bonding

What is a covalent bond?

Covalent bonding is a sharing of a pair of electrons between two atoms. Simple yet complicated, the covalent bond gives us many compounds and structures in chemistry. However, before the discussion on covalent bonding starts, an explanation on drawing chemical structures.

What are Lewis Dot Structures?

G.N. Lewis developed a method to visually determine covalent bonds. Called electron dot structures or Lewis Dot Structures. This method is based on an old hypothesis on bonding but works well in explaining covalent bonds and structure of molecules consisting of non-metals. This technique is included in this text; however, it will not work for transition metals.

One of the principles behind electronic dot structures is that an "electron shell" will contain eight electrons. This electron shell is now considered the s and p-orbitals of the highest energy level of the atom. The electron dot diagram for an atom is the picture of the number of valence electrons for the atom. By convention, a dot is used to signify the electron, but a cross or an x can also be used. The maximum number of valance electrons the atom can have is eight, thus the "shell" is closed. The diagram starts with the elemental symbol in a middle of an imagined square. Place a dot for each valence electron starting on one side of the box and move clockwise. Do not pair up the electrons until all four sides have an electron. Figure 6.1 shows the electron dot structure for lithium, carbon, and phosphorous.

Figure 6.1: Examples of Lewis Dot Structures of Atoms

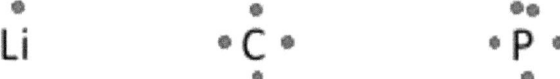

What are the exceptions to the electron dot structures?

There are a few exceptions to the electron dot structures. The first exception is that hydrogen and helium can have a maximum of two electrons in the shell. This makes sense because these elements only have the 1s orbital available for bonding. The second exception is that boron and aluminum can be stable with just six electrons around it. The last exception is that sulfur and phosphorous can have 10 or 12 electrons in their electron shell. This is due to the complexity of the bonding orbitals in the molecules.

How to create an electron dot structure for a molecule?

There are several methods used to create an electron dot structure for a molecule. The one that is presented in this text is the best method, in my opinion. However, it is not the only one out there, and if the reader finds another method that is applicable to them, feel free to use it. In this explanation, electrons will be designated by dots, and a line signifies a bond, two shared electrons, between two atoms.

How to draw a covalent bond using Lewis Dot Structures?

There are several methods to draw covalent bonds using Lewis dot structures. This is the one I use when working on these compounds:

1. Obtain the chemical formula of the compound and sum the number of valence electrons the molecule has. If the molecule has a charge, add or subtract the appropriate number of electrons.

2. Place the symbol of the left most element in the center, then draw the other atoms around the center like a box.

    a. If there is more than one element that should be in the center, these atoms are connected.

3. Use a dash or a single line to represent a single bond and connect each atom with one line.

    a. The atoms must be next to each other

4. Count the number of electrons used in bonding. Each line is equal to 2 electrons.

5. Place the electrons around the most electronegative atoms first, up to eight, then the center atom.

6. Each atom should have 8 electrons that it owns or shares. (I usually say that each atom "sees" eight electrons.)

7. If an atom does not have eight, move an electron pair from the atom next to it so it can share more than two electrons – this is called making a double bond.

8. Some molecules have an odd number of electrons. The atoms in the center usually have the odd electron that is unpaired. This type of molecule is called a radical and is very reactive.

Examples of Lewis Dot structures

Below in Figure 6.2 are examples of Lewis dot structures, starting with the simplest CH_4 or methane. Next example is Cl_2CO that shows a double bond, then NO_2 that shows the radical form. A radical molecule is a molecule that has an unpaired electron present in the molecular orbital. These compounds are highly reactive.

Figure 6.2: Example of Lewis dot structures

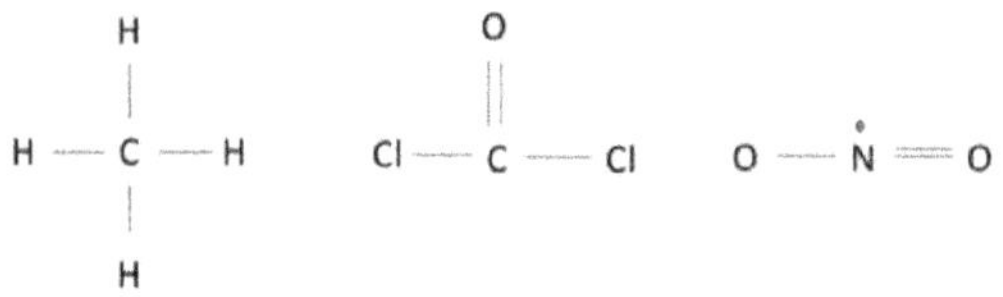

```
      H                   O
      |                   ||
H ─── C ─── H     Cl ─── C ─── Cl     O ─── N ═══ O
      |
      H
```

What is covalent bonding?

Covalent bonding is a type of bonding where two electrons are shared between two atoms. This type of bonding is typically found when two nonmetals join.   Unlike the ionic bonding described in chapter 7, covalent bonds are very directional; the paired electrons can be found in a certain volume of space, usually between the two atoms sharing the electrons.

How do atoms share electrons?

When two atoms share electrons, the atomic orbital of the unpaired electrons must overlap.   The electronic orbital must have similar energies, not equal but similar. There are several different types of overlap; the common type is when an s-orbital interacts with another s-orbital or a p-orbital. The direct overlap is described as a sigma bond (σ-bond).   When the two orbitals mix, two molecular orbitals are formed, a bonding orbital which is lower in energy and an antibonding orbital which is higher in energy. The average energy of the molecular orbitals is the average of the two atomic orbitals before mixing. Figure 6.3 shows this overlap. Other factors that affect the covalent bonding is that the energy levels of the orbitals need to be similar and the electronegativity of the atoms determines the polarity of the bond. The resulting interaction is called a covalent bond and the atomic orbitals are now changed to molecular orbitals.

Figure 6.3: Overlap of two s orbitals making a sigma bond

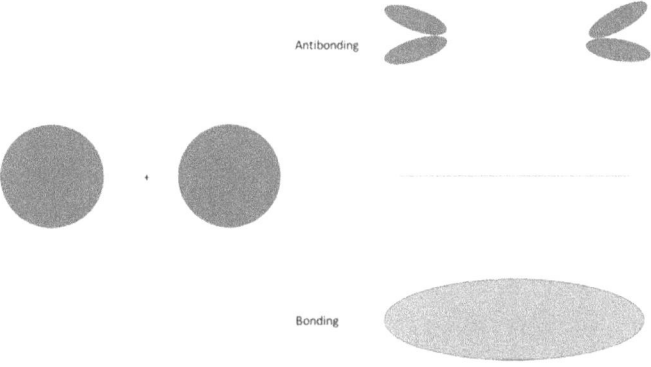

What are double and triple bonds?

A double bond is a combination of a sigma and a pi bond. The electron pairs are located between the atoms, acting like a cushion of negative charge than draws the atoms together, creating a shorter bond. A triple bond consists of a sigma bond and two pi bonds, with the electrons pulling the atoms closer together.

Interaction of two identical atoms

When two nonmetal atoms form a molecule, the atomic orbitals are the same energy and there must be an unpaired electron in each orbital. The two orbitals overlap and mix to form two molecular orbitals. One is lower in energy and the other is higher in energy. The lower energy orbital is called the bonding orbital and the higher energy orbital is called the anti-bonding orbital. Notice that the result of bond formation is the release of energy. In this type of molecule this is a common result.

To break the bond, energy is required to move, or excite, the electron from the bonding orbital to the antibonding orbital, breaking the bond. Figure 6.4 shows the energy diagram for the formation of H_2 from two hydrogen atoms.

Figure 6.4: Energy diagram for the formation of hydrogen gas

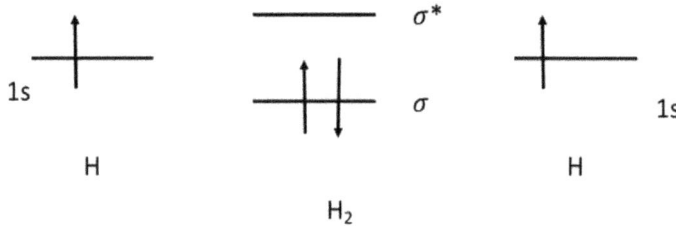

What is bond order?

Bond order is the number of bonds a pair of atoms have, usually one, two, or three. But sometimes you can have a bond order of 1/2, 3/2 etc. The bond order is calculated by counting the electrons in the bonding orbitals, subtracting the number of electrons in the antibonding orbitals, then dividing by two. Notice that the orbitals below the valence orbitals cancel each other out. Focus on the highest energy orbitals and count the electrons in those orbitals.

How do you know that this theory is correct?

If we look at the formation of oxygen gas, the Lewis dot structures indicate that it has two lone pairs and two double bonds. If the molecular orbitals are used, as shown in Figure 6.5, it is obvious that the oxygen molecule contains two unpaired electrons. These unpaired electrons give the molecule the magnetic properties and reactivity.

Figure 6.5: Molecular orbital diagram of O_2

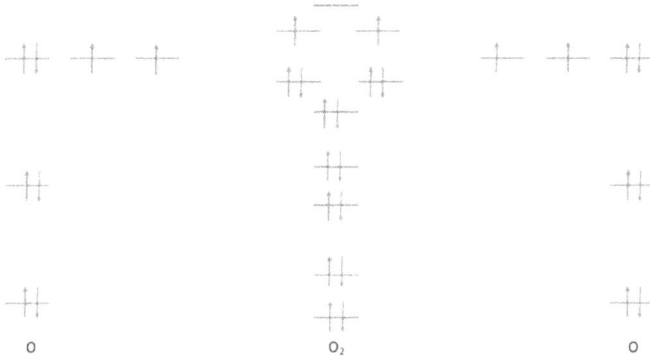

How does the overlap between two different atoms occur?

When atoms of two different elements interact, several actions occur. To form the bond, an electron in one of the atoms must be excited. Then the orientation of the atomic orbital must be correct for the orbital to overlap. To overlap, the atoms must collide with each other. Finally, when the orbitals overlap and form a molecular orbital, energy is released when the electrons fall into the bonding orbital.

How is energy used to form bonds?

There is a type of chart used to show reaction energies. The x axis is called the reaction coordinate, which basically indicates time on a relative scale. The y axis is the energy of the system, usually in J/mol. For most reactions energy is required to start the reaction, then enough energy is released that the reaction continues. This is called an exothermic reaction, illustrated in Figure 6.6

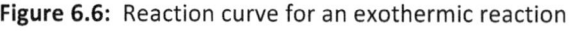

Figure 6.6: Reaction curve for an exothermic reaction

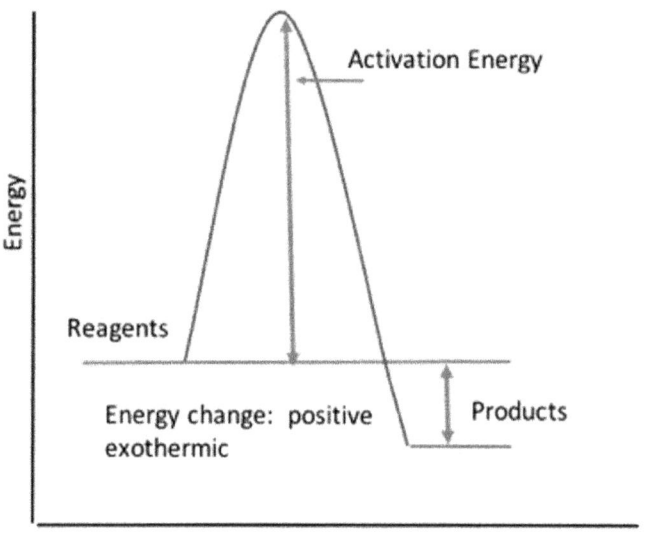

Reaction Coordinate

Some reactions absorb energy from the surroundings. As shown in Figure 6.7, the system's energy level is higher at the end. This type of reaction is called an endothermic reaction. On both diagrams, the difference between the starting level and the top of the curve is called the activation energy. The reaction will not start until the energy of the system reaches that point. The system will lose energy until the reaction has finished. When the reaction is complete, the final energy level will indicate if the reaction is exothermic or endothermic.

Figure 6.7: Reaction curve for an endothermic reaction

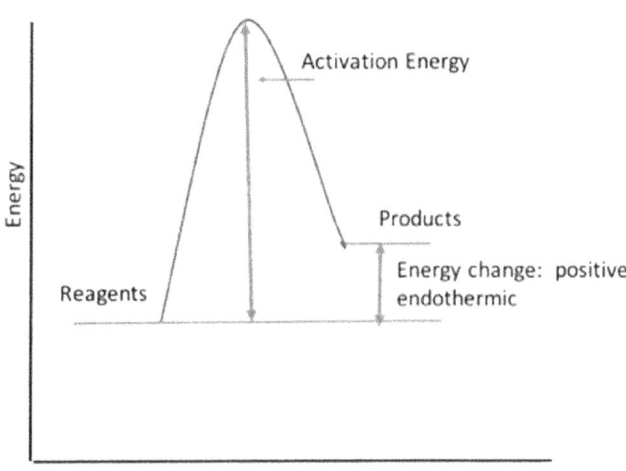

Reaction Coordinate

Molecules containing more than two atoms.

The molecular orbitals of molecules containing more than two orbitals are complicated. In a comparison between the Lewis dot structure for carbon and the electron configuration, the number of unpaired electrons is not the same. Both are correct. Carbon can interact with four other atoms to form covalent bonds and the electronic structure of the atom indicates the presence of two unpaired electrons. To deal with this difference, chemists have hypothesized that the atomic orbital of the carbon atom will mix to form hybrid orbitals. The atomic orbitals involved in mixing are the valance orbitals, the s and p orbitals of the highest energy level. The designation given to these orbitals is the letter and number of orbitals involved in the mixing. The most common hybridization is the sp^3 hybrid orbital. The 3 p orbitals and the s orbital mix to form 4 sp^3 orbitals.

What are molecular shapes?

With the hybridization of atomic orbitals, the molecules can form specialized three dimensional structures. Each type of hybridization will have a core structure. In sp^3 hybrid orbitals the molecule will form structures based on a tetrahedron. These sp^3 orbitals will have either lone pairs or sigma bonds. If the central atom has 4 sigma bonds, then the geometry is a tetrahedron. When one of the bonds is replaced with an electron pair, then the geometry is trigonal pyramidal. When the central atom has two sigma bonds and two lone pairs, the geometry is bent.

When the hybridization is sp^2, the unhybridized p orbital is either empty, as with Al and B, or is used in a pi or double bond. This geometry is called trigonal planar. If one of the sigma bonds is replaced with an electron pair, then the geometry is called bent. An sp hybridization has two p orbitals unhybridized. These orbitals are used to form two double bonds or a triple bond. There is only one type of geometry for this hybridization, the linear molecule.

What are bond angles?

A bond angle is the angle between two atoms attached to a single central atom. The bond angle is dependent on the hybridization of the central atom and the presence of lone pairs. Per the valance shell electron pair repulsion theory (VSEPR), the electronic regions will move as far away from each other as physically possible. An electron region consists of any bond and electron pairs.

What is resonance?

Resonance is the concept of a double bond moving two different atoms with one end anchored at a central atom. This concept was developed as an explanation for having several Lewis structures that were correct yet different. Now the resonance structures can be explained by a molecular orbital that shares two electrons with more than two atoms that are bonded together. Chemists illustrate this concept by drawing a dashed line between the central atom and the atoms involved in the resonance structure. Figure 6.8 shows the nitrate anion resonance structure.

Figure 6.8: Resonance structure of nitrate anion

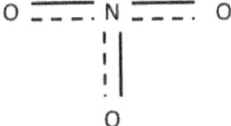

What are polar bonds and molecules?

Earlier in the text electronegativity was discussed. This is the amount of pull the atom has on electrons in a covalent bond. Some atoms have a stronger influence on electrons than others. So, when two atoms of vastly different electronegativity are covalently bonded, the sharing of the electrons in the bond is uneven and the bond has a slightly negative end at the atom with the highest electronegativity, and a slightly positive end at the atom with the lowest electronegativity. Figure 6.9 shows a schematic of a polar bond.

Figure 6.9: Schematic of a polar bond

A molecule is polar if the molecule contains just one polar bond or if the multiple polar bonds do not cancel each other out. The concept of symmetry and multiple polar bond in a molecule will be left for a more advanced text.

How to write and name covalently bonded compounds
When writing chemistry reports or texts, the compounds are written in a "code" to reduce the amount of space on a page. For example, CH_4 is shorter than writing methane or "one carbon and four hydrogens". A systematic method to write the formulas is used. The method to name and write covalently bonded compounds is a relatively simple task for the compounds that have two or three elements in the molecules.

The formula is written by determining the number of atoms of each element in the compound. Then the most electropositive, or the element to the left or below the others, is listed first. The number of atoms of that element is placed in the lower right as a subscript, with number one being implied. The most electronegative element is written next with the number of atoms listed as a subscript.

How to write or call the compounds

The naming of compounds is very systematic. In chemistry, there are three systems that you will learn as budding chemist. Two of these, ionic and simple covalent nomenclature, will be described in this text. First are the simple covalent compounds. The first element is named after the element. H is hydrogen and P is phosphorous and so on. The elemental name of the last atom is used with the suffix of -ide. (Table 6.1) A Greek prefix is used to denote the number of atoms of that element in the compound. (Table 6.2 lists the prefixes used in this system). The ionic nomenclature system will be described in chapter 7.

Table 6.1: Common Elemental Anions

Element	Anion	Name of the anion
Hydrogen	H^-	Hydride
Fluorine	F^-	Fluoride
Chlorine	Cl^-	Chloride
Bromine	Br^-	Bromide
Iodine	I^-	Iodide
Oxygen	O^{2-}	Oxide
Sulfur	S^{2-}	Sulfide
Nitrogen	N^{3-}	Nitride
Phosphorous	P^{3-}	Phosphide
Carbon	C^{4-}	Carbide

Table 6.2: Prefixes Used in Naming Simple Molecules

1: mono -	5: penta -	9: nona -
2: di -	6: hexa -	10: deca -
3: tri -	7: hepta -	
4: tetra -	8: octa -	

Examples

CO_2 - carbon dioxide. Mono is understood in this case

CO - carbon monoxide. Mono must be used in this example since it describes the number of oxygen and not to be confused with carbon dioxide.

P_2O_5 - diphosphorous pentoxide

Chapter 7: Ionic Bonding

What is ionic bonding?

Ionic bonding is described as an electrostatic attraction between a positively charged atom or molecule with a negatively charged atom or molecule. When matter becomes charged, the substance will either gain or lose electrons. A positively charged atom or molecule is called a cation – pronounced as "kat-i-on". A negatively charged atom or molecule is called an anion. There are two ways a cation is formed; one by the addition of energy to remove the electron and the second is by reacting with another element that will take the electron. There are two ways to form an anion. One is to assume that the electron is "free" or unattached to any atom and the atom will absorb the electron and release energy. The second method is that the atom will take the electrons from another element. For this chapter, we will use the first interaction exclusively.

How are cations formed?

Cations are positively charged atoms or molecules. Cations are formed by adding energy to the atom, and the electron will absorb enough energy to escape the electrical forces of the nucleus. A typical gas phase reaction of the removal of an electron is described below.

$$Li + energy \rightarrow Li^+ + e^-$$

This energy is defined as the ionization energy. Any energy above the ionization energy is transferred to the electron as kinetic energy. All reactions will involve a transfer of energy. With this example, energy is absorbed by lithium, which goes to the electron. The electron is then excited to another energy level and is ejected from the atom, forming a cation.

Is it possible to remove all the electrons from an atom?

There is a limit to the number of electrons that can be taken from the atom. The electrons that can be taken from the atom are from the highest occupied energy level. Therefore, since lithium has only one electron in the second energy level, only one electron can be removed. The amount of energy between the first and second levels is quite large and it is difficult to remove the electron in the first level chemically.

All metals will form cations and some will even have the ability for form more than one cation. This will depend on the electron configuration and energy absorbed by the atom. There are few molecules that form cations, the only one that will be addressed in this chapter is ammonium cation (NH_4^+).

What are the assumptions we can make with cations?

There are a couple of rules that will help understand which elements have only one form of the cation. All elements in the first column (alkali metals) will only form a cation with a plus one charge. Hydrogen is an exception to this rule, due to its electron configuration. It will form a +1 cation but it will also absorb an electron to form a -1 anion. Hydrogen has only the 1s orbital in the ground state and it is partially filled. Hydrogen can lose the electron to form H^{+1} or it can gain an electron to form the H^{-1} anion.

Elements in the second column (alkali earth metals) will form a +2 cation only. The +1 cation cannot be formed, it is unstable and it doesn't take much energy for the second electron to leave. The electron configurations of these elements helps explain why these elements will only form a +2 cation. These elements have two electrons in the highest energy level; both are in the s orbitals. The p orbitals in these elements are considered to be virtual, so adding electrons is unrealistic. Removing these electrons is easier. Once the first electron is removed, the second electron is removed quickly.

A few more elements can form a single cation. The explanation on why these occur will be left for another text. These elements are boron, aluminum, silver, and zinc. The rest of the metallic elements can form more than one cation, and most of these elements are in the transition metal portion of the periodic table.

How are anions formed?

Anions are formed by absorbing a free electron or an electron from another atom and giving off energy as the electron reduces the amount of energy it has. An example of this process is shown below:

$$Cl + e^- \rightarrow Cl^- + energy$$

With this reaction, energy is released. Remember, the electron must be ionized from another metal so it has energy. When the atom accepts the electron, the excess energy is released as a form of heat.

What are the assumptions with anions?

As with cations, the anions have a set of rules to make things easy to remember. All elements in column 17, the halides, will form -1 anion.

Oxygen and sulfur will form -2 anions, and nitrogen and phosphorous will form -3 anions. Please remember that this is for elemental anions and that most molecular anions have oxygen attached to the nonmetal. These anions may differ from one another by the number of oxygens bound to the nonmetal, but the charge on the anion remains constant.

How do ionic bonds form?

Ionic bonds are formed when the ions of opposite charges interact with each other. For this chapter, we will assume that the cations and anions have already been formed. The ionic bond itself is omnidirectional, the attraction of the ions occurs in every direction. One other fact is that the formula of the ionic compound must be neutral, or have no charge. The total positive charge must equal the total negative charge. When the solid is formed, a solitary unit cannot be isolated. This is due to the nature of the ionic bond.

How can an ionic bond break?

The energy associated with ionic bonds is dependent on the charge of the ion and the distance away the ions are. It takes a lot of heat to melt a salt and even more energy to make the compound a vapor. However, polar compounds, like water and methanol for example, can use their molecules to break the bond and dissociate the salt. The result is a change in the energy, usually heat is produced, although some salt will absorb heat.

What are salts?

Salt is a general name for an ionic compound that has a cation and an anion. It could be a simple as sodium chloride, $NaCl$, or as complex as clay. Physical properties of salts range from white crystals to crystals of various colors. They have high melting and boiling points, but are easily dissolved in polar solvents, especially water. These compounds will also increase the conductivity of any polar solvent when dissociated in the liquid.

How to form salts?

When sodium cations and chloride ions are mixed together, they will bond together to form a solid crystal. To write the formula of this crystal, write the small ratio of ions that can exist and remain neutral. Sodium is a +1 cation and chloride is a -1 anion, therefore the formula is NaCl. These types of salts are characterized as 1:1 salts.

How to name simple salts?

There is a simple scheme to name inorganic (or ionic) compounds. Each salt is comprised of two components, a cation and an anion. The cation name is unchanged from the element's name. For example, Na^+ is called the sodium cation or just sodium when used in a name.   Naming the anion is a little different, because the ending of the name will indicate if it is an anion or the type of anion.   For most nonmetal elements; the name of the anion is the elemental name with the last two or three letters removed and replaced with the suffix –ide.

What are molecular anions?

Molecular anions are those molecules that can exist with a charge. A molecule is two or more nonmetals bound together in a covalent bond, which was discussed in the previous chapter. There is only one common molecular cation, ammonium cation, which has the formula of NH_4^+. All other common molecular anions are negatively charged. Table 7.1 list the common anions, including the elemental anions.

Table 7.1: Common anions

Fluoride (F^-)	Chloride (Cl^-)	Bromide (Br^-)	Iodide (I^-)
Oxide (O^{2-})	Sulfide (S^{2-})	Arsenide (As^{2-})	
Nitride (N^{3-})	Phosphide (P^{3-})		
Nitrate (NO_3^-)	Nitrite (NO_2^-)		
Phosphate (PO_4^{-3})	Hydrogen Phosphate (HPO_4^{2-})	Dihydrogen phosphate ($H_2PO_4^-$)	
Phosphite (PO_3^{3-})	Hydrogen phosphite (HPO_3^{-2})	Dihydrogen phosphite ($H_2PO_3^-$)	
Carbonate (CO_3^{2-})	Hydrogen carbonate (HCO_3^-)	Cyanide (CN^-)	
Hydroxide (OH^-)			
Hypochlorite (ClO^-)	Chlorite (ClO_2^-)	Chlorate (ClO_3^-)	Perchlorate (ClO_4^-)

Why are the molecular ions named?

The names for the molecular ions originated from a previous nomenclature system. These names have been used for so long it was difficult to change them, so the chemists adapted. The anions can be separated into different groups based on the central atom. To distinguish the different anions a system indicating the number of oxygen was developed. The suffix –ate indicates that the anion has a higher number of oxygen in the family. The suffix –ite indicates the anion has a lower number of oxygen in the family. For those families with four members, the hypo- prefix means less and the per- prefix indicates more.

How do you write the formulas of these compounds?
To write the formula, determine the ions in the compound and list the cations first, then the anions. For most compounds, there is only one cation. If the anion is molecular, for example the nitrate ion, then the anion formula must remain as a subunit of the formula. If you have more than one unit of the molecular anion, the formula of the anion must be in parenthesis.

How do you write the names of these compounds?
The name of the compound is the cation followed by the anion. The simple salts usually have one formula that is neutral, making the name easy. If the cation is a transition metal, the cation must have the charge of the charge listed in parenthesis and as a roman numeral. This will differentiate between salts of two or more different metal cations of the sample element.

Special System – Acids and Bases
Acids and bases have a separate system of nomenclature, mainly because of the historical and common usage of these compounds. An acid is a compound that produces a proton when dissociated in water. If the anion ends in the suffix –ate, then the name of the acid is anion + -ic acid. If the anion ends in -ite, then the name is the anion plus -ous acid. All elemental or mineral acids are named by using hydro plus anion plus -ic acid. For example, HCl is hydrochloric acid. Bases are

those compounds that produce a hydroxide anion when dissociated in water.

Chapter 8: Metallic Bonding, Crystal Structures, and Bond Strength

What are metallic bonds?

Metals have a specific type of bonding that is very difficult to describe properly in any introductory textbook. The analogy that many authors use for a metal bond is a nucleus in a sea of electrons. This analogy is descriptive but not accurate with the current hypothesis and theories of atomic orbitals.

How do metallic bonds form?

When metals form a solid, the atoms are packed together in a crystal structure that provides the maximum number of atoms in a volume and minimizing the amount of repulsion from neighboring atoms. This crystal structure varies depending on the atom size and electronic structure. When metal atoms are bonding with each other the atomic orbitals interact with the orbitals from other metallic atoms in a three-dimensional array. The orbitals involved are usually the s and p orbitals of the valence shell and the d orbitals of the element. These orbitals overlap each other whether if the orbital contains electrons or not.

The combination of atomic orbitals of different atoms form an interconnecting band of molecular orbitals that stretch from one end of the material to another. It is in this interconnecting band of orbitals that the electrons can travel through the metal. Now, we have a hypothesis on how scientists can explain the conductivity, malleability, and other physical properties of metals.

What are crystal structures?

Many compounds, both metallic and nonmetal, form crystal structures when a solid. Some of these structures are quite simple while others are more complex. This book will review the crystal structures that the elements will form and leave the complex structures for inorganic chemistry or geology.

Cubic Structures

There are three types of crystal structures based on the cube; simple, body centered, and face centered cubic. The simple cubic is a cube with atoms located at each corner of the cube. The body-centered cubic structure is similar to the simple cubic with an atom on each corner of the cube; there is also an atom in the middle of the cube. The face-centered cubic is very crowded. There are atoms on each corner of the cube and an atom in each side of the cube. Figure 8.1 shows each of the three structures.

Figure 8.1: Three simple cubic crystal structures

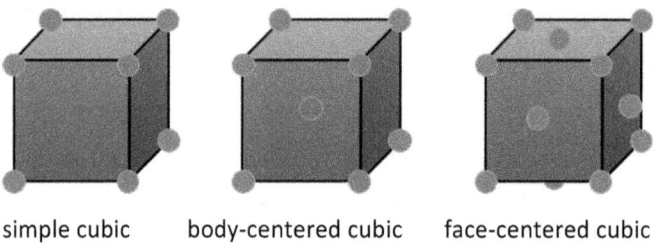

simple cubic   body-centered cubic   face-centered cubic

What is meant by closest packed crystal?

The crystal structure that is the most efficient in packing the atoms as close as possible is described as hexagonal packed structure. This structure is where you have three hexagons on top of each other. There are two common closest packed structures with the elements. One is a cubic structure – the face centered cubic. The other is called hexagonal closest packed. The hexagonal packed structure is ABA, while the face-cubic structure is ABC. Figure 8.2 shows the differences between the two structures.

Figure 8.2: Hexagonal packed crystal structure

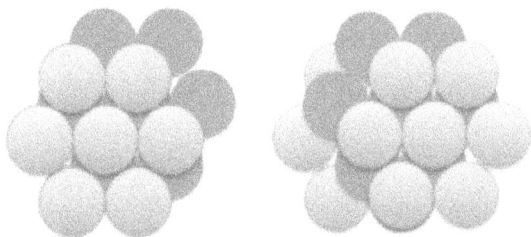

In the hexagonal packed (ABA) crystal structure, note that in the second image the first and third levels are overlapping each other.

With the face-centered cubic (ABC) crystal structure, the second and third layers overlapped the holes made by the first layer. See the second image in figure 8.3 for details

Figure 8.3: Face-centered cubic crystal structure

What are the other elemental crystal structures?

There are four more crystal formations that the elements form when in the solid phase. These structures are uncommon and only several elements form these crystal structures. The first one is the rhombohedral, Figure 8.4, which is the crystal structure for mercury, bismuth, arsenic and antimony. In this structure, all sides have the same length but the angles are equivalent but not 90 degrees.

Figure 8.4: Rhombohedral crystal structure

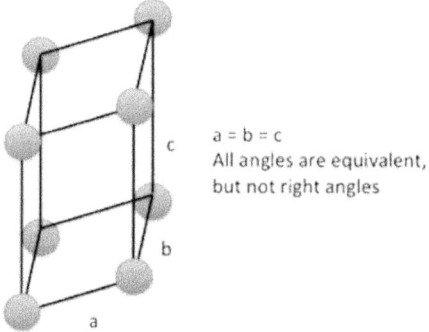

a = b = c
All angles are equivalent,
but not right angles

Phosphorous, polonium, and plutonium each will form the monoclinic crystal structure. The monoclinic structure can be described as a square rod, Figure 8.5.

Figure 8.5: Monoclinic crystal structure

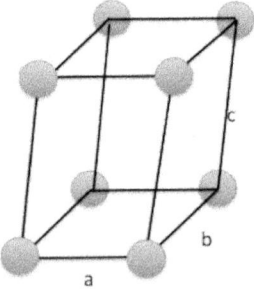

a = b ≠ c
All angles are right angles,
except for the angle between a and c

The third structure is orthorhombic, Figure 8.6. This crystal structure has sides of various lengths, but all the angles are right angles. Sulfur and all halogens will form crystals of this type.

Figure 8.6: Orthorhombic crystal structure

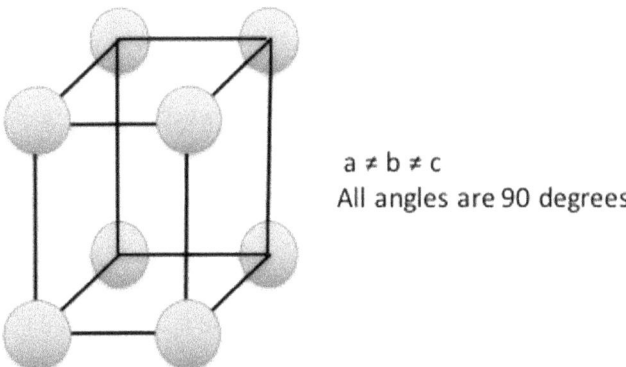

$a \neq b \neq c$
All angles are 90 degrees

The last crystal structure is the tetragonal, Figure 8.7. Most of the sides are equal and all the angles are right angles. This structure is uncommon, only indium and tin share this crystal structure.

Figure 8.7: Tetragonal crystal structure

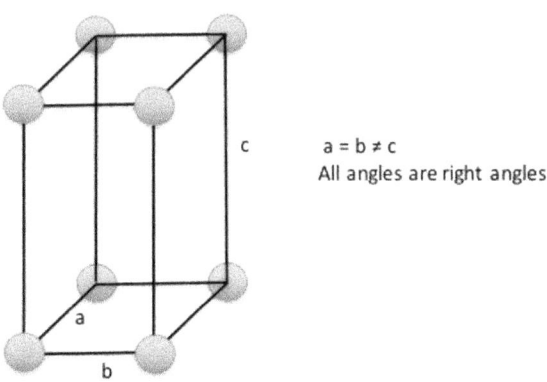

$a = b \neq c$
All angles are right angles

How can these crystal structures be used to calculate atomic radii?

The atomic radii are calculated from crystal structures. The distance between two corners of the crystal is measured and depending on the crystal structure, the radii can be calculated. Below are the common calculations for the determination of atomic radii.

For simple cubic and rhombohedral structures, measure the distance between the two corners and divide by two. Since these structures have the corner atoms touching each other, the measurement is simple. For the body centered cubic structure, the distance from one corner to the opposite corner, through the center of the atom, is taken. Then this measurement is divided by 4 to obtain the atomic radii. The tetragonal and the monoclinic crystals, the atomic radii is calculated by measuring the distance between the two corners of the shortest side and dividing by two.

Is there a relationship between crystal structure and electrical conductivity?

Electrical conductivity depends on the electronegativity of the element and the crystal structure. Generally going from left to right, the electronegativity increases and the electrical conductivity decreases. There is a definite change in electrical conductivity as the elements become a nonmetal. Going down the column, the crystal structure does influence the electrical conductivity. If the crystal structure changes from a closest packed structure to one with more spaces, the electrical conductivity decreases.

What is an alloy?

An alloy is a mixture of two or more metallic elements that are bonded together with a metallic bond. Alloys are also considered a mixture, where the elements in small amounts are dissolved in another metal. Steel is one common example of an alloy. In steel, iron is the major component and carbon is one of the additives. Carbon can fit into the vacant holes or spaces in the iron crystal structure, or lattice, and strengthens the iron so it is harder and less reactive to oxygen. With many alloys, the crystal structure is based on the major component of the alloy with the minor components mixed irregularly in the lattice.

Which bond is the strongest?

Bond strength is a complex discussion. First, criteria on quantitating bond strength must be established. As an example, we can use the melting point as a determination of bond strength. Using this criterion, ionic bonds should be classified as the strongest since many compounds with ionic bond have a high melting point. However, chemically, these compounds can dissociate in water and other polar solvents, whereas covalent bonds are resistant to this type of reaction.

Metallic bonds are strong, according to melting points, and are resistant to most chemical reactions. Most metals are prone to reactions with H^+. Since metallic and ionic bonds are three-dimensional bonds or the bonds are omni-directional, this makes the substances prone to reactions by other chemicals.

What about the covalent bonds? These bonds are localized and highly directional. The electrons are located between the atoms involved in the bond. Energetically, the sigma bond has the lowest energy between the sigma and pi bonds. However, to break apart two atoms, you need more energy to break a triple bond than a single bond.

Is it possible to break a sigma bond first, then a pi bond?

This question is a good one because of the molecular orbital energies. If a photon of light containing energy can excite the electron in the sigma bond to the anti-bonding orbital, the electron in the pi orbital will automatically move down to the empty hole, releasing energy. The net energy absorbed will be equal to the energy needed to excite the electron in the pi orbital, even though we began this question by exciting the electron in the sigma orbital. Figure 8.8 shows this process by energy diagrams of the orbitals.

Figure 8.8: Excitation of the σ electron

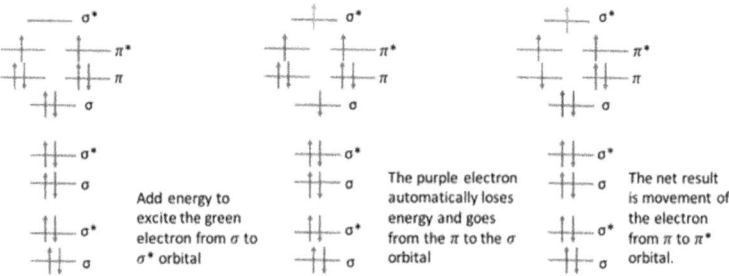

A couple of notes concerning this example. The electron from the σ orbital can be excited to either π* or σ* orbitals. The net energy absorbed is the energy needed to excite the π electron to the antibonding orbital.

Chapter 9: The Mole

Why use the mole?

Before the text launches into a full-scale explanation of chemical reactions, the student needs to be familiar with the concept of indirect counting. Indirect counting is a skill that originated with carpentry and other fields that dealt with purchasing large quantities of an item.   If the weight of 100 nails is known, let's say 1 kg, then the carpenter can order 10 kg of nail and he can be assured that he will purchase approximately 1000 nails, give or take 1 or 2 nails. Chemists have taken this concept and have used it extensively.

How to use dimensional analysis

This chapter will use a technique called dimensional analysis. Every number that we use in this chapter and in the rest of the text will have a unit attached to it. The number is multiplied by the unit, but it does not change the measurement or size of the number. Rather, it tells the reader or user what the number means. For example, if I have a number 10 and I need to tell the reader how far I walked, there is no meaning with the number so the reader can infer that I could have travelled 10 kilometers while another reader could infer that I meant to say I travelled 10 days. The units are multiplied, so when the given data requires a different unit, we can cancel the old unit and replace it with the new one.

What are conversion factors?

Conversion factors are a ratio of relatable measurements or calculations. Used to convert a number from one measurement system to another, chemists have used conversions factors to determine the amount of product a reaction can deliver or how much reagents are needed. An example of a conversion factor is the conversion of one inch to centimeters. A fact is that 1 inch = 2.54 centimeters. To make this a conversion factor, divide 2.54 cm by 1 in. There are two ways to write this factor, depending on which unit to keep:

$$\frac{2.54 \text{ cm}}{1 \text{ in}} \text{ or } \frac{1 \text{ in}}{2.54 \text{ cm}} = 1$$

Since both ratios are equal to one, the actual measurement does not change, but the descriptor does and the magnitude has to change with the unit.

What is a mole?

A mole is the name of a unit of counting. For example, 100 is called a hundred and 1,000,000 is called a million. People have used names for numbers for a very long time and as a society are fairly used to this concept. Not all names are based on a system of 10. Time has a system of names for numbers based on a system of 60, then 24. In chemistry, the unit of mole is equivalent to 6.02×10^{23} items. This is a lot of items. We can never count that amount in our lifetime, especially if we have an item such as cars or buses. 6.02×10^{23} is also called Avogadro's number, after the scientist who proposed it. This is a huge number and as the reader continues through this course, it will be apparent that chemists must handle large numbers.

How do chemists use the mole?

Chemists use the ideal of a mole as a unit of counting and mass. Chemists weigh a compound, which we can determine the number of moles based on the substance's molecular mass. But first, lets perform a few simple calculations.

How to convert number of molecules to moles?

We can convert the number of atoms, molecules or electrons to moles by using this conversion factor: 1 mole = 6.02×10^{23}. Remember, the mole is a term to communicate a large number to a reader or listener.

$$\left(\frac{5.25 \times 10^{15} \ atoms \ Cu}{1}\right) * \left(\frac{1 \ mole \ Cu}{6.02 \times 10^{23} \ atoms \ Cu}\right) = 8.72 \times 10^{-9} \ mole \ Cu$$

In the example above, Avogadro's number has the unit of atoms of Cu, which is the unit of the number we are transforming. This is a trait of Avogadro's number; the unit of the number is the unit of the number being transformed. This can cause students headache as they try to understand the importance of this calculation.

Of course, everyone should be able to convert moles of a substance back to atoms or molecules:

$$\left(\frac{1.40 \ moles \ H_2O}{1}\right)\left(\frac{6.02 \times 10^{23} \ molecules \ H_2O}{1 \ mole \ H_2O}\right) = 8.43 \times 10^{23} \ molecules \ H_2O$$

Please note that moles are not the same as molecules. These are two different units of measurement.

How to calculate the atomic mass in grams?

In chapter 2, the mass of a proton and a neutron was determined to be roughly 1 Da. For hydrogen, the average atomic mass is 1.01 Da. Now, since a Dalton is very tiny, we can calculate the mass of 1 mole of hydrogen by the example below. Use the mass of the atom in Daltons and multiply by the conversion factor from Daltons to kilograms. Then multiple by Avogadro's number and finally convert the kilograms to grams.

$$\left(\frac{1.01 \text{ Da of H}}{1 \text{ atom of H}}\right)\left(\frac{1.66 \times 10^{-27} \text{ kg}}{1 \text{ Da}}\right)\left(\frac{6.02 \times 10^{23} \text{ atom H}}{1 \text{ mole of H}}\right)\left(\frac{1000 \text{ g}}{1 \text{ kg}}\right) = \frac{1.01 \text{ g of H}}{1 \text{ mol H}}$$

This calculation system is the same, no matter if we have the mass of an atom or a protein in Daltons. 1 mol of H = 1.01 g of H is also called the molar mass of the element.

How can this relate to the periodic table?
The average atomic mass listed for each element in the periodic table can be interpreted in two ways. If we want to know the mass of the atom, the table will provide the value in Daltons. If looking for the mass of a mole of an element, then the value in the table has the unit of gram per mole. The neat ideal on the periodic table is that the magnitude of the number can be either Daltons or g/mol, depending on the use. This is one reason why the periodic table never has the units for mass.

Calculating the mass of an element given the moles.
If you know the number of moles of an element, for example, 2.5 moles of silver, then it is simple to calculate the mass of silver in your hand. An example calculation is presented below:

$$\left(\frac{2.5 \text{ mole Ag}}{1}\right)\left(\frac{107.87 \text{ g Ag}}{1 \text{ mol Ag}}\right) = 269.68 \text{ g of Ag}$$

Of course, reversing the equation and calculating the moles of an element when the mass is given is another important technique. An example calculation is present below where you are given 200 g of carbon and the number of moles is required for the reaction:

$$\left(\frac{200 \text{ g of carbon}}{1}\right)\left(\frac{1 \text{ mole C}}{12.01 \text{ g C}}\right) = 16.65 \text{ g of C}$$

How to determine the molar mass of a compound?

If the formula of a compound is given, it is simple to calculate the molar mass of the compound. It is the sum of the average mass of each atom in the molecule. For example, water (H_2O) has two hydrogen atoms and an oxygen atom in the molecule, or a mole of water has two moles of hydrogen and one mole of oxygen. The formula can be stated either way. The mass of a mole of water is 2.02 g (from hydrogen) plus 16.00 g (from oxygen) for a total of 18.02 g of water per mole of water.

How to determine the moles of a compound

Starting with a mass of a compound, for example 20 g of CaO, we can determine the moles of the compound by dividing the mass of the compound by its molecular weight. For these problems, it is inferred that the reader must calculate the molecular or formula weight first:

Ca =   40.08

O =   16.00

Molecular weight = 56.08 g/mol

$$\left(\frac{20 \text{ g CaO}}{1}\right)\left(\frac{1 \text{ mol CaO}}{56.08 \text{ g CaO}}\right) = 0.357 \text{ mol CaO}$$

Of course, as a chemist it is expected that you can perform the calculation in reverse. How much does 0.500 mol of CaO weigh?

$$\left(\frac{0.500 \text{ mol CaO}}{1}\right)\left(\frac{56.08 \text{ g CaO}}{1 \text{ mol CaO}}\right) = 28.04 \text{ g CaO}$$

What is the information that a chemical formula contains?

Given the chemical formula of a compound, any chemist can determine the moles of the element. For example, a chemist has 1.5 moles of $K_2Cr_2O_7$, we can calculate the number of moles of chromium by the following example:

$$\left(\frac{1.5 \text{ moles of } K_2Cr_2O_7}{1}\right)\left(\frac{2 \text{ mole Cr}}{1 \text{ mole } K_2Cr_2O_7}\right) = 3.0 \text{ moles of Cr}$$

A simple activity that most chemistry students undergo is the determination of the mass percentage of an element in a compound. Using $K_2Cr_2O_7$, the mass percentage of chromium can be determined.

Example:

Units can be either Da/atom or g/mole

2 K	2 * 39.01 = 78.02
2 Cr	2 * 52.00 = 104.00
7 O	7 * 16.00 = 112.00

total   294.02 ← this is the molar mass

mass percent of K	(78.02/294.02) * 100 = 26.54%
mass percent of Cr	(104.00/294.02) * 100 = 35.37%
mass percent of O	(112.00/294.02) * 100 = 38.09%

total percent   = 100%

Sometimes the percentage may not total 100%, this is due to rounding errors. If the total percentage is not between 99 and 101%, then an error is in the calculation or the data given in the problem.

How can the empirical formula be determined?

The empirical formula is the ratio of atoms of each element in the compound or molecule. Use the lowest common denominator and ensure that the number of atoms are whole numbers – no fractions allowed. For example, the empirical formula of sodium chloride is NaCl, one atom of sodium and one of chlorine. If we write Na_2Cl_2, this will be incorrect since we can divide the number of both sodium and chlorine atoms by two. Many molecules share the same empirical formula, especially with organic chemistry.

To calculate the empirical formula, the data presented is a result of a combustion experiment. This is when a sample of your compound is burned in a pure oxygen atmosphere and the resulting products are quantitated. The final result is a list of the percent composition of the elements in your compound. For this chapter, the focus is on the results and how to determine the empirical formulas.

The method to determine empirical formulas is the following:
1. Convert the mass of each element to the percent composition in the compound
   a. often, the problem provides the percent composition, then go to step 2
2. Assume that there is 100 g of the compound, this will convert the percentage to grams
3. Convert the mass of each element to moles of an element
4. Find the lowest number of moles and divide the moles of each element by that number
5. The lowest number of moles will become 1

6. If one of the other numbers is close to 0.5, then multiple all the values by 2

7. If the values within 0.1 to a whole number, then round to that whole number

There are many other ways to determine empirical formulas, plus some short cuts. As long you can achieve the same answer the method is valid.

For our example, 5.60 g of calcium hydroxide was analyzed and it was found to contain 3.03 g of calcium, 2.42 g of oxygen, and 0.15 g of hydrogen. To calculate the empirical formula, determine the percent composition first.

$$\frac{3.93 \ g \ Ca}{5.60 \ g \ sample} * 100\% = 54.1\%$$

$$\frac{1.57 \ g \ O}{5.60 \ g \ sample} * 100\% = 43.2\%$$

$$\frac{0.10 \ g \ H}{5.60 \ g \ sample} * 100\% = 2.7\%$$

Then – assume we have 100g of sample and we have

54.1 g of calcium, 43.2 g of oxygen, and 2.7 g of hydrogen.

Next we convert these masses to moles.

$$\frac{54.1 \ g \ Ca}{1} * \frac{1 \ mol \ Ca}{40.08 \ g \ Ca} = 1.35 \ moles \ Ca$$

$$\frac{43.2 \ g \ O}{1} * \frac{1 \ mole \ O}{16.0 \ g \ O} = 2.70 \ moles \ O$$

$$\frac{2.70 \ g \ H}{1} * \frac{1 \ mole \ H}{1.01 \ g \ H} = 2.67 \ moles \ H$$

Divide each by the lowest number of moles, which is 1.35

Ca = 1.0, O = 1, and H = 1.98 or 2

Therefore, the empirical formula is CaO_2H_2. A person can skip the percent composition step and use the mass of each element that was measured. It will be quicker, but the percent composition step can tell the chemist if the molecule is missing an element.

How to convert empirical formulas to molecular formulas

To obtain the molecular formula of a substance, both the empirical formula and the molecular weight must be known. As shown above, once the percent composition is known, the empirical formula is a set of simple calculations. The determination of the molecular formula is simple since the molecular formula is an integral multiple of the empirical formula. Find the mass of the empirical formula, called the empirical mass, and divide it into the molecular mass, the integer that is found is multiplied through the empirical formula to find the molecular formula.

An example is the determination of the molecular formula for benzene. The empirical formula for benzene is CH, the empirical mass is 13.02 g/mol. The molecular mass is 78.11 g/mole

78.11/13.02 = 6.00

So, the molecular mass is 6 times the empirical mass, then benzene consists of 6 units of CH. The molecular formula of benzene is C_6H_6.

Chapter 10: Stoichiometry and Calculations

Stoichiometry is a term that describes a process of calculating mass of chemicals using a balancing chemical equation. A balanced chemical equation is a chemical reaction which the number of atoms in the reactants are equal to the number of atoms in the products.

How to balance chemical reactions?

One of responsibilities of a chemist when performing a reaction is to use just enough reagents to make the amount of product. Thus, making sure the reaction equation is correct and balanced is essential. A balanced reaction means that the chemical equation that describes the reaction has identical number of atoms on either side of the reaction arrow. In other words, the number of atoms as reagents must equal to the number of atoms as products.

Steps to Balance Chemical Reactions:
1. Write the reaction
2. Count the number of atoms for each element as a reactant and as a product
3. Start with the heaviest element first, balance the number of atoms on each side by adding molecules that contain the element
4. Repeat step three with the next heaviest element until hydrogen and oxygen are left
5. If hydrogen and oxygen are still unbalanced, work on hydrogen then oxygen

a. If oxygen has an odd number on one side, the other side may use ½ of an oxygen gas molecule to balance the reaction

6. If a fraction is present in the reaction, multiply the reaction by the denominator of the fraction – usually 2

7. Check to see if the number of elements are equal on each side of the reaction

Example:

Chromium (III) nitrate reacts with sodium sulfate to produce chromium (III) sulfate and sodium nitrate.

Unbalanced equation is

$$Cr(NO_3)_3 + Na_2SO_4 \rightarrow Cr_2(SO_4)_3 + NaNO_3$$

count the atoms of each element

element	Left side	right side
Cr	1	2
Na	2	1
N	3	1
S	1	3
O	13	15

Balance the equation - starting with the heaviest element first

The reaction has 1 Cr on the left and 2 on the right, therefore I need to add one additional $Cr(NO_3)_3$ to the left side. This also changes the count of nitrogen and oxygen.

Unbalanced equation is

$$2\ Cr(NO_3)_3 + Na_2SO_4 \rightarrow Cr_2(SO_4)_3 + NaNO_3$$

count the atoms of each element

element	Left side	right side
Cr	2	2
Na	2	1
N	6	1
S	1	3
O	22	15

Next, balance the sulfur. There is 1 sulfur atom on the left and 3 on the right. Add two more Na_2SO_4 to the left for a total of 3 Na_2SO_4. This also changes the number of sodium and oxygen atoms.

Unbalanced equation is

$$2\ Cr(NO_3)_3 + 3\ Na_2SO_4 \rightarrow Cr_2(SO_4)_3 + NaNO_3$$

count the atoms of each element

element	Left side	right side
Cr	2	2
Na	6	1
N	6	1
S	3	3
O	30	15

Next, balance the sodium atoms. There are six sodium atoms on the left side and one on the right. Add 5 more $NaNO_3$ to the right side to have a total of 6. This will change the number of nitrogen and oxygen atoms as well.

Unbalanced equation is

$$2\ Cr(NO_3)_3 + 3\ Na_2SO_4 \ \rightarrow\ Cr_2(SO_4)_3 + 6\ NaNO_3$$

count the atoms of each element

element	Left side	right side
Cr	2	2
Na	6	**6**
N	6	**6**
S	3	3
O	30	**30**

Now, the chemical equation is balanced.

$$2\ Cr(NO_3)_3 + 3\ Na_2SO_4 \ \rightarrow\ Cr_2(SO_4)_3 + 6\ NaNO_3$$

What information can be gleaned from a balanced reaction?
The balanced reaction provides the reader with a lot of information. The coefficient informs us of how much reagents we need, how much will be made and if a reagent limits the reaction. All that is required is one piece of information, mainly the mass or amount of a chemical involved in the reaction. One assumption that is made with these problems is the reaction occurs as written, there are no side reactions and incomplete reactions taking place. These concerns are addressed in other texts and college classes.

What is a net ionic equation?

Once the reaction equation is balanced, then chemists can reduce the equation to an expression that involves the components that is involved in the reaction. This reduced form of the reaction equation is used to express single displacement and double displacement reactions.   Table 10-1 provides a list of symbols used to designate the of the substance. If the reaction is in aqueous solution, usually a solid or a gas is produced. One to two of the components are not involved in the reaction – these are called spectator ions. Others are involved in the reaction.

Table 10-1: Phase symbols

(s) – solid	(l) – liquid
(g) – gas	(aq) – aqueous

In the example below, the reaction of lead nitrate with sodium iodide occurs as written below:

$$Pb(NO_3)_{2(aq)} + \ 2\,NaI_{(aq)} \ \rightarrow \ PbI_{2(s)} + 2\,NaNO_{3(aq)}$$

Note the state of the compounds, (aq) indicates that the compound is dissolved in water, (s) means that the compound forms a solid and precipitates out of solution to the bottom of the beaker. If one of the products is a gas, then (g) is used. For a liquid that is produced, usually water, the (l) is used.

To determine the spectator ions, the reaction is separated into ions. That is, all compounds that have (aq) by the formula can be separated into the appropriate ions. As shown below, the charged ions have the aqueous state included on the chemical symbol. Often, for ions, aqueous solution is assumed and the (aq) is not added to the formula.

$$Pb^{2+}_{(aq)} + 2\,NO^-_{3(aq)} + 2\,Na^+_{(aq)} + 2\,I^-_{(aq)} \rightarrow PbI_{2(s)} + 2\,Na^+_{(aq)} + 2\,NO^-_{3(aq)}$$

Any product that forms a solid, liquid, or gas is kept together. Then the ions that appear on each side of the equation are called the spectator ions and can be removed from the equation. What is left is called the net ionic equation and this equation describes the ions that are involved in the reaction. For the example reaction, sodium iodide can be replaced with potassium iodide or calcium iodide. The cation is not important in the reaction, only the anion.

$$Pb^{2+}_{(aq)} + 2\,I^-_{(aq)} \rightarrow PbI_{2(s)}$$

The calculation to determine mass of product.

For this type of question, you are given the reaction and the mass of one of the reagents. Sometimes, you may be given the moles of one reagent. Either way, to need the balanced equation to convert the moles of reagent to moles of product, then find the mass of product. One assumption with this type of problem is that all the other reagents are in excess. The example of this type of problem is we have 5 g of lead nitrate in solution and it reacts with sodium iodide to form lead iodide. The question is how much lead iodide will we make?

1. Start with the balanced equation. If not balanced, you need to balance it.

    $$Pb(NO_3)_{2(aq)} + 2\,NaI_{(aq)} \rightarrow 2\,NaNO_{3(aq)} + PbI_{2(s)}$$

2. Determine the molecular/formula weight of the substances in the question

    a. Formula weight of $Pb(NO_3)_2$ is 331.2 g/mol

    b. Formula weight of PbI_2 is 461.0 g/mol

3. Convert the mass of reagent to moles, then use the ratio of PbI_2 to $Pb(NO_3)_2$ to convert moles of $Pb(NO_3)_2$ to PbI_2, then calculate the mass of the product.

$$\left(\frac{5\,g\,Pb(NO_3)_2}{1}\right)\left(\frac{1\,mol\,Pb(NO_3)_2}{331.2\,g\,Pb(NO_3)_2}\right)\left(\frac{1\,mol\,PbI_2}{1\,mol\,Pb(NO_3)_2}\right)\left(\frac{461.0\,g\,PbI_2}{1\,mol\,PbI_2}\right)=6.96\,g\,PbI_2$$

How to calculate the mass of a reagent

Sometimes, the amount of each reagent must be known before you conduct the reaction. Especially if the reagents are expensive. Using the lead iodide reaction above, the amount of sodium iodide can be calculate using the following format.

4. instead of using the ratio of PbI_2 to $Pb(NO_3)_2$, use the ratio of $Pb(NO_3)_2$ to NaI.

$$\left(\frac{5\,g\,Pb(NO_3)_2}{1}\right)\left(\frac{1\,mol\,Pb(NO_3)_2}{331.2\,g\,Pb(NO_3)_2}\right)\left(\frac{2\,mol\,NaI}{1\,mol\,Pb(NO_3)_2}\right)\left(\frac{149.9\,g\,NaI}{1\,mol\,NaI}\right)=4.53\,g\,NaI$$

How to determine the amount of a reagent from mass of a product

If you know how much product you need to make, you can calculate the amount of reagents you need to add. Using the reaction in the previous examples, if 50.0 grams of lead (II) iodide is needed, we can calculate the amount of sodium iodide need for the reaction. Starting with lead (II) iodide, we convert the mass to moles and then convert the mole of lead (II) iodide to moles of sodium iodide.

$$\left(\frac{50.0\ g\ PbI_2}{1}\right)\left(\frac{1\ mol\ PbI_2}{461.0\ g\ PbI_2}\right)\left(\frac{2\ mol\ NaI}{1\ mol\ PbI_2}\right)\left(\frac{149.9\ g\ NaI}{1\ mol\ NaI}\right) = 32.5\ g\ NaI$$

How to determine the limiting reagent of a reaction

This type of problem can be difficult to solve, especially if the reaction is complex. The data you are given in this class of problem is the amount, either in grams or moles, of the reagents and the chemical equation. This type of problem occurs in everyday life, from a sandwich maker determining how many sandwiches he can make to a baker in a factory calculating how much ingredients he needs to purchase. There are many ways to solve this problem, the steps below comprise the method I use:

1. Convert the amount of reagents to moles, if necessary
2. Make sure the chemical equation is balanced, if not, balance it.
3. Using the moles of each reagent, determine the mole of a product that is created. For this step, pick one product and keep with that compound throughout this example.
4. Compare the moles of product made by each reaction, the reagent that made the least amount of a product is the limiting reagent.
5. From this step, you should be able to calculate the mass of product that will be made plus any leftover reagents.

Example

A chemist has 5.00 g of chromium (III) nitrate in a solution and 5.00 g of sodium sulfate in a solution. When the two solutions are mixed, the reactants will react according to the equation below. Which reagent is limiting?

$$2\ Cr(NO_3)_3 + 3\ Na_2SO_4 \rightarrow Cr_2(SO_4)_3 + 6\ NaNO_3$$

1. The chemical equation above is balanced
2. The next step is to calculate the molecular mass of the compounds
   a. Chromium (III) nitrate is 238.01 g/mol
   b. Sodium sulfate is 142.0 g/mol
3. Determine the mass of a product. It doesn't matter which one you choose, so we will pick sodium nitrate.

$$\left(\frac{5.00\ g\ Cr(NO_3)_3}{1}\right)\left(\frac{1\ mol\ Cr(NO_3)_3}{238.01\ g\ Cr(NO_3)_3}\right)\left(\frac{6\ mol\ NaNO_3}{2\ mol\ Cr(NO_3)_3}\right)\left(\frac{85.00\ g\ NaNO_3}{1\ mol\ NaNO_3}\right)$$

$$= 5.35\ g\ of\ NaNO_3$$

$$\left(\frac{5.00\ g\ Na_2SO_4}{1}\right)\left(\frac{1\ mol\ Na_2SO_4}{142.0\ g\ Na_2SO_4}\right)\left(\frac{6\ mol\ NaNO_3}{3\ mol\ Na_2SO_4}\right)\left(\frac{85.00\ g\ NaNO_3}{1\ mol\ NaNO_3}\right)$$

$$= 5.99\ g\ of\ NaNO_3$$

You can stop at moles of product; it may be easier for students to understand if mass is calculated.

4. Compare the two results, the reagent that produced less of the product is the limiting reagent, in this case chromium (III) nitrate.

What is percent yield?

The percent yield is a simple calculation that describes the efficiency of a reaction. It is calculated by taking the mass of a product that was produced and dividing by the theoretical yield – which is the mass of the product that you should receive if the reaction was complete. The theoretical yield is calculated by determining the limiting reagent and how much product is made with the limiting reagent.

A sample calculation is described below. A reaction to produce aspirin is supposed to make 35 g of product, but only 29.8 g of aspirin was recovered, what is the percent yield?

The theoretical yield is 35g.

$$\frac{29.8 \; g \; of \; aspirin \; made}{35 \; g \; of aspirin-theoretical} * 100\% = 85.1\% \; yield$$